Praise for *Brand Power Built In*

"Think of every product you love—the brand is inseparable from the offering. Lifang's practical guide shows you how to create these connections and build remarkable experiences."

—Colin Raney, CEO and Cofounder, Ray

"Lifang reframes brand as a strategic engine—integral to product and demand. It's a must-read for leaders building resilient, customer-centric organizations to drive long-term value."

—Marc-Antoine Jarry, Managing Director and Global Head of Marketing and Communications Strategy, Accenture

"Lifang has written the book I wish existed when I started leading brands in tech—hard-won wisdom too important to keep to herself. What makes it extraordinary is who she is: brilliant, generous, and unafraid to challenge convention."

—Sophie Ho, Head of Brand Marketing, Claude at Anthropic

"Too many startups leave massive value on the table by treating brand as a luxury for big companies. Lifang delivers a powerful lesson for anyone building something new: make brand core to product development."

—Matteo Vianello, Executive Creative Director and Cofounder, Squero

"This book offers essential frameworks for integrating brand thinking early in product design to drive go-to-market success. It's an invaluable resource for students and professionals alike."

—Ben Lee, Professor, USC Annenberg School for Communication and Journalism

BRAND POWER BUILT IN

How Tech Products *Really* Win Hearts and Minds

Lifang He

Matt Holt Books
An Imprint of BenBella Books, Inc.
Dallas, TX

Matt Holt is an imprint of BenBella Books, Inc.
8080 N. Central Expressway
Suite 1700
Dallas, TX 75206
benbellabooks.com
Send feedback to feedback@benbellabooks.com

Matt Holt and *BenBella* are federally registered trademarks.

Printed in the United States of America
10 9 8 7 6 5 4 3 2 1

Library of Congress Control Number: 2025046895
ISBN 9781637748527 (hardcover)
ISBN 9781637748534 (electronic)

Editing by Lydia Choi
Copyediting by Michael Fedison
Proofreading by Rebecca Maines and Sarah Vostok
Indexing by Elise Hess
Text design and composition by Aaron Edmiston
Cover design by Matteo Vianello
Printed by Lake Book Manufacturing

*To all builders at heart—those who pursue excellence,
seek meaning, and dare to lead differently.*

CONTENTS

Introduction   1

PART I: THE OPPORTUNITY

CHAPTER 1   Rethinking Brand for Tech Products   13

CHAPTER 2   Building In Brand Power from the Start   31

PART II: BUILD

CHAPTER 3   Develop Your Positioning Strategy   51

CHAPTER 4   Establish Your Brand Identity   75

CHAPTER 5   Integrate Your Brand into Your Product Experience   95

PART III: LAUNCH, SCALE, AND ITERATE

CHAPTER 6   Launch with a Sequenced Go-to-Market Strategy   119

CHAPTER 7   Scale Your Brand to Accelerate Acquisition and Retention   143

CHAPTER 8   Iterate After Measuring Impact   169

PART IV: SUPPORT A BRAND-POWER CULTURE

CHAPTER 9   Cultivate a Culture That Powers Brand Success   191

Conclusion: Putting It All Together   211
Glossary of Terms   219
Acknowledgments   225
Notes   229
Index   240

Introduction

"Brand" Is a Dirty Word in Tech

In 2022, I remember saying to a new hire at Amazon: "Brand is a dirty word in the tech industry. People don't appreciate it. Let's not use the word 'brand.'"

This was ironic, since at the time my job title was Head of Brand Marketing. And I had spent my entire career up to that point building brands.

But this thought had been on my mind for a few years. The truth is, many tech companies misunderstand the power of a strong brand. They try to build products first and brands later, treating them as two separate things. Too often, brand is seen as something fluffy—just a visual identity, an aesthetic layer, or a beautiful wrapper. None of that is actually true.

When Ellen Redefined Amazon Key

In 2017, Amazon launched a new service called Amazon Key In-Home Delivery. I wasn't working for Amazon yet, but I thought it was impressive that the company could now offer users a secure way to let delivery drivers into their homes so their packages wouldn't get stolen. But most Amazon customers weren't that enthusiastic. Many were skeptical about this new offering or found it downright creepy.

Comedians loved it.

For instance, Ellen DeGeneres made fun of the new service on her popular TV talk show:

> You order something on Amazon, and you grant them access to your house through a cloud—what could possibly go wrong there? And then they drop off your package, rifle through your medicine cabinet, and leave. That's what happens. The reason Amazon is doing this is because a lot of people are worried that [if] their packages . . . [are] left outside, someone will steal it. So the solution to that is to let people into your house, where they have the opportunity to steal everything in your house.[1]

Ouch!

By the time I joined Amazon in 2018, the In-Home Delivery service was struggling. The media reviews were mixed, few customers were signing up, and even those who did sign up weren't using it frequently. With too few customers, operational costs per customer were way too high. I was brought in to lead marketing and brand for Amazon Key and help turn around the business.

This was the context for a meeting I had with a product and engineering lead. He was a literal tech genius who had started working on a new form of remote access control to solve his personal parking dilemma[2] in Seattle. But he had no use case in mind for his cool innovation, which could enable secure deliveries behind a customer's gate or inside multi-unit buildings. Our mission was to integrate these upgrades into Amazon Key to make the service more appealing, especially for business customers.

But no matter how much we explored different positioning and narrative approaches, such as framing it as a unified access control system, we kept hitting a brick wall. *Amazon Key already had an unappealing brand.* By developing the product first, without considering the brand, we had gotten stuck in a corner.

Two years later, in 2020, Amazon quietly phased out the In-Home Delivery service and shifted to other secure delivery solutions that customers felt more comfortable with.

Seeking a Better Way

Amazon Key wasn't the first time I had faced this challenge. I'd often found myself in rooms dominated by product managers and engineers, discussing how to launch a new product and market its features to customers. Once again, they were treating brand as an afterthought. In the product development process, the cool tech came first, while the needs or desires of the customer were not considered systematically. Brand was often misunderstood as just a logo, or something for big companies to worry about, not for new products. Therefore, it wasn't prioritized.

Because of this misunderstanding, marketers have to come up with strategies to help sell new products after the fact. Too often, product builders jump into building the products, features, and technology and don't spend enough time on the why and the purpose of the product until it becomes too late.

This is a prevalent, industry-wide challenge. We've witnessed the costly consequences of neglecting customer needs, value, and experiences. Each year, over 30,000 new consumer products are launched, yet more than 90%[3] end in failure. Humane's Ai Pin, for instance, was a promising wearable tech product aiming to shape the intelligence age. Even though it was founded by two longtime Apple veterans with sufficient funding, the product failed to deliver real customer value or meaningful experiences, ultimately shutting down in February 2025—less than two years after launch, wasting millions in investment and years of work.

Winning customers is a common goal, of course. Every company wants to win. Every company claims to be customer-centric. But creating powerful connections—the kind of experiences that build strong customer relationships and make your product irreplaceable in their minds—is incredibly difficult. It requires not only understanding customer problems but also designing holistic solutions from the outset—embedding customer connections into the product's DNA and building great experiences around it.

This is what brand is *really* all about—creating meaningful connections and experiences through the product and everything around it to win hearts and minds. But this is where most companies fail—because their brand thinking either does not exist at all or remains superficial. For example, in many cases, brand positioning, the most foundational strategic work to inform execution, is outsourced. That's the missed opportunity and untapped value most companies overlook.

••

I knew there had to be a better way.

Earlier in my career, I had the privilege of working on one of the most iconic products of our time, the iPhone. As part of the team responsible for its global launch, I saw firsthand how Apple's success was built on more than just great technology. The true power lies in a deep sense of care for customers and the seamless integration of brand and product to deliver a delightful experience. Every detail, from the keynote launch event to the product experience to the marketing and advertising campaigns, was deliberately crafted to reinforce the product positioning and build strong relationships with customers.

As Steve Jobs once said, "It is in Apple's DNA that technology alone is not enough—that it's technology married with liberal arts, married with the humanities, that yields us the result that makes our heart sing."[4] This encapsulates Apple's core brand difference and its philosophy on product innovation, design, and customer experience, creating products that are not just functional but emotionally resonant.

Once at Amazon, I brought my brand-forward mindset to an environment that operated like a collective of startups. Driven by its relentless innovation, speed, and scale, I had to adapt my approach to fit a different culture and operating model. Through experimentation and perseverance, I found my way to integrate brand thinking into early-stage product development. I collaborated with some visionary founding teams to build and scale several high-growth startup projects, including Ring, Prime benefits expansion, and Amazon Pharmacy.

Throughout this journey, I sought resources on product development, brand building, and go-to-market strategies for new products, but I couldn't find a comprehensive, up-to-date guide that addressed the interdisciplinary challenges I experienced. This isn't just a process challenge of bringing people earlier into product development. It's a fundamental business and product strategy gap that many don't even realize they have.

Over the years, former coworkers, industry peers, and graduate students have reached out to me with similar challenges: whether tech companies care about brand, where to start and invest, what to prioritize, how to collaborate with product and user experience (UX) teams, how to connect brand with performance marketing, how to measure success, and how to earn a seat at the table. These were the same questions I had when I started at Amazon.

Even when startups hire brand experts, they often don't fully understand what brand really means or how it drives value. As a result, those experts often find themselves justifying their roles or educating internal stakeholders on why it matters, rather than doing the strategic work they were brought in to do.

This persistent knowledge gap, the lack of common language and shared principles between the product, marketing, and business communities, along with the transformative shifts I've observed in the industry, inspired me to create a practical handbook—one that I wish had existed for myself, my colleagues, and industry friends as we navigated these challenges. I believe a book is a scalable way to share knowledge and hard-won lessons with those facing similar challenges, providing a comprehensive and integrated resource to address this complex topic.

What Is *Brand Power Built In*?

Brand Power Built In is a repeatable framework for embedding brand power and customer connection into a product's DNA from day one—not as an afterthought, but as a built-in competitive advantage. Instead of treating brand efforts as a fluffy downstream marketing activity that follows product development, this approach brings brand strategy upstream as a powerful source of value creation. Your brand thus informs and integrates with product vision and business strategy right from the start.

Your brand is the heart of your products. Just as the heart gives life to the body, your brand gives meaning and purpose to your products. It shapes every decision about what you build and how customers experience it. This creates customer perceptions and relationships that become inseparable from your product itself.

This approach is rooted in three principles.

First, *product experience is the most powerful brand builder*. If you apply brand strategy to shape your products from the start, you will build better products that naturally connect with customers and lay a strong foundation for go-to-market success and sustainable growth.

Second, *your product is the entire customer journey*. Your concept of "product" needs to be defined more broadly. Every touchpoint, from your product naming, packaging, and website to your acquisition and onboarding experience, impacts customer perceptions and product success. If you expand your definition of product, you will build better experiences across the entire customer journey.

Third, *brand building is a shared responsibility*. Brand building is no longer just a marketing activity—it becomes a cross-functional responsibility across product, UX, marketing, and anyone else involved in shaping customer experiences. While this may sound simple, it often challenges how most companies operate. Embracing this mindset shifts how companies build and launch new products, transforming the entire product development and go-to-market process and how teams work together.

This is an approach that I've seen work throughout my career at some of the most successful tech companies. By elevating brand investment upstream during the building phase, you can achieve stronger customer engagement, faster market adoption, and outsized growth.

Brand Power Built In is not Marketing 101 or Product 101. It is a practical, interdisciplinary playbook, designed for product builders, marketers, and business leaders alike. It will help anyone who aspires to drive innovation and growth to navigate the journey of creating better products, cohesive customer experiences, and more valuable businesses.

Why Now?

Of course, building products with brand power is not the only path to starting a business. Some companies start with interesting technology, while others focus on solving a specific customer problem. Not every founder wants to take a holistic approach to design products and businesses. But, regardless of the path you take, the core principles we explore in this book can enhance your products and create better customer experiences.

Today, many companies—particularly in emerging spaces like AI—face a similar challenge: pouring massive investments into developing new products and services without clear customer use cases or brand strategies. As the speed of innovation and the product development cycle accelerates, you simply can't afford to miss the most important foundations that help maximize your potential and success.

This book aims to answer critical questions facing many product builders and marketers, including:

- How do you build brand power and customer connection into product development?
- How do you go to market to establish a strong position for a new product?
- How do you transform a new product into a sustainable and successful business?

To gain broader perspectives and provide insights applicable to a wide range of companies and business stages, I interviewed dozens of founders, product leaders, and marketing executives. This book distills the learnings and insights I've gained from those conversations and throughout my career. It also incorporates success and failure stories from a new generation of leaders and brands, including Ring, Rivian, PillPack, Notion, Robinhood, Airbnb, Figma, and many others.

How This Book Is Organized

Part I shows the huge opportunity open to you when you build in brand power from the start of your product development process.

Part II explores the three key steps of developing and integrating your positioning strategy, brand identity, and product experience.

Part III shows how to launch, scale, and iterate new products for maximum impact.

Finally, Part IV reveals how to cultivate a culture that will support brand power going forward.

Along the way, we'll explore many case studies as well as the common pitfalls of each stage in this process. You will learn where to focus your energy and investment to build meaningful customer relationships throughout your journey.

This book is designed in a modular way to address the most pressing challenges in each product phase. You can read it straight through in order, or jump around between chapters to focus on your most urgent problems first.

Every successful product began as an idea, but turning that idea into something extraordinary isn't about finding a silver bullet—it's about making thoughtful decisions at every step.

Whether you're leading a startup or driving innovation within an established company, your opportunity is clear: When you build brand power from the start, you don't just build better products—you also win hearts and minds and create a more valuable business. This book is your guide to help you get there, no matter where you are in your journey.

PART I

The Opportunity

Why should founders, product builders, and marketers care about brand? Most tech ventures fail to stand out or connect meaningfully with customers until it's too late. What separates promising products from true market leaders isn't just great technology—it's embedding brand power into your product's DNA from day one, *not* as an afterthought after the product is built. That's the hidden opportunity most companies miss.

This part reveals the true power of brand for tech products. You'll learn why defining and integrating brand strategy from the earliest stages helps you avoid costly mistakes and how it can become a force for product differentiation, customer connection, and business growth. This opportunity will inspire you to rethink your product development and go-to-market strategy to start cultivating your brand power from the start.

Chapter 1

Rethinking Brand for Tech Products

From Brand Is Fluffy to Brand Is How You Win

In tech, brand is often misunderstood. Too many founders and product builders dismiss it as a fluffy marketing activity that can wait. But this assumption kills promising products. This chapter explores what we get wrong, what we miss, and why we must rethink our current approach to building products.

From Doorbot to Ring: Winning with Brand Power from Day One

Do you remember the story of Ring? The company that was rejected on *Shark Tank* was later acquired by Amazon for over $1 billion.

For decades, the traditional home security market was dominated by established companies like ADT, whose services were expensive and inflexible. Customers had little choice but to commit to multi-year contracts. Ring took a radically different approach and changed this stagnant industry.

In 2011, Jamie Siminoff, a serial inventor and entrepreneur, created the first Wi-Fi video doorbell in his garage. Initially called "Doorbot," Siminoff's invention debuted on ABC's *Shark Tank* in 2013, seeking $700,000 for a 10% stake in the company. "The doorbell has not changed since it was invented in 1880—until now," Siminoff told the sharks. "Introducing the Doorbot, the first-ever video doorbell built for the smartphone. With Doorbot, you can see and speak with visitors from anywhere."[1]

While the pitch piqued interest, Mark Cuban, one of the sharks, rejected him. "I like it. I think you're going to do great with it. But . . . when I jump in, I've got to add enough value that this company . . . worth $7 million, could be worth $80 million, $90 million. I just don't see that progression, and for that reason, I'm out." Three other sharks also didn't see the potential and passed.

Despite this very public rejection, Siminoff continued to refine his idea. The Doorbot was technically innovative but lacked a compelling story and brand identity. Initially named project "f5_v001,"[2] the prototype reflected its tech-focused origins. To realize its full potential, Siminoff sought to transform the product into a brand with a powerful vision. He and Simon Cassels, who later became the company's chief marketing officer, reimagined the future of Doorbot. Together, they created a new brand called Ring, which would embody their vision of connected home monitoring and community security. It started with a simple idea to change the doorbell, but they quickly realized its bigger potential.

The breakthrough came from a key insight: Most household traffic centers around the front door, the first point of contact with your home, your family, and your home security. As their original vision deck stated, "Your home security starts right in your doorbell on your doorstep right where you need it most."[3] This realization expanded their thinking beyond positioning the company around a simple doorbell alternative. If home security starts with the front door, it could extend to the entire home and throughout the neighborhood. Ring's vision and future product roadmap were born, with the doorbell as their first entry point into the home security space.

In 2014, the Ring Video Doorbell was officially introduced. "The name Ring has a couple of meanings," Siminoff explained. "It's the sound a doorbell makes, but it also comes from the ring of security we create around your home and, in time, your community."[4] The tagline "Always Home" captured the essence of the brand and became the foundation for Ring's product roadmap and ongoing innovation.

Since launching its first doorbell, Ring has expanded into an ecosystem of home monitoring and security products and services, including indoor and outdoor cameras, alarm systems, smart lighting, subscription plans, and the Neighbors app. Over time, the company has introduced more than 50 devices,[5] forever changing the home security category and how people connect with their homes and neighborhoods.

Reflecting on the early days of building Ring, Cassels shared, "The ideas and narrative from [f5_v001] formed the basis of our external communication strategy, enabling us to effectively convey to the world what Ring represented. Importantly, this unified vision rallied our investors and supporters, ultimately facilitating our acquisition by Amazon."[6] Nearly a decade later, Cassels proudly noted that almost everything laid out in the early vision deck eventually became a reality.

In 2018, Ring was acquired by Amazon for over $1 billion.[7] Dave Limp, former senior vice president of Amazon Devices, said to the Ring team, "We acquired you because you were a brand."[8] In 2025, Ring became one of the largest home security companies and one of the most popular home security brands, competing fiercely with traditional security systems market leader ADT for 94 million potential households in the United States.[9]

Consider this: If Jamie Siminoff had continued to use the name Doorbot and hadn't focused on building a brand in the early days, would the company have achieved the same success? It's unlikely that Doorbot would have become a global home security brand. Trying to rebrand after selling millions of units would have been too late, diminishing its value and potentially requiring more investment to reenter the market. Doorbot would have struggled to grow its brand equity, expand product lines, and attract customers, severely limiting its growth potential.

The story of Ring provides critical lessons on how early brand efforts created significant business value for new ventures. Ring differentiated itself from traditional home security services by establishing its positioning and narrative early on, then building a compelling brand identity with a name that resonated with customers. This foundation helped Ring grow from a garage startup to a billion-dollar acquisition, eventually becoming one of the largest home security companies today.

Three Myths on Why Tech Companies Undervalue Brand

Despite clear evidence that brand power drives business value as we see from Ring's journey, many tech companies still don't fully appreciate the value of brand efforts.

What are we—the tech community as a whole—missing?

Here are three myths that hold back even the most successful tech companies, shaping how we think about and invest in brand building and product development. These myths influence daily decisions about resources, priorities, and process and how we collaborate with each other.

Myth #1: "Brand is separate from product."

Myth #2: "Product innovation alone will drive success."

Myth #3: "We can build a brand through advertising."

Let's unpack each myth and how they directly impact business outcomes.

Myth #1: "Brand Is Separate from Product."

One of the most persistent myths is the disconnection between product and brand. This common misperception causes companies to overlook brand efforts during early-stage product development or go-to-market planning, undermining the potential to shape product experiences and foster customer connections early on.

If you ask 10 people what brand means, you might get 10 different answers. For some, it's limited to logo, awareness, or advertising. For others, it's about creative work and aesthetics. Terms like "brand," "branding," and "brand building" are often used interchangeably.

This ambiguity and misunderstanding come from several factors: a lack of shared vocabulary and clarity about what brand truly means and what value it brings to products and services, historical contexts that shaped our interpretation, and the unique dynamics of the tech industry. Consequently, teams aren't always aligned when making business decisions or collaborating on brand initiatives.

What Is Brand?

The concept of "brand" existed as early as 2700 BC, when livestock owners used a mark on their cattle to prove ownership and differentiation. This marking served as an identity and quality assurance for the product.

Thousands of years later, the fundamental purpose of a brand remains unchanged. Just like those merchants marked their goods to distinguish quality, your brand gives your products and services a distinctive identity, sets you apart in a crowded market, and builds trust and emotional connections with your customers. The difference is that how we express a brand has expanded far beyond its original marking on cattle.

I know how confusing it can be to talk about brand. People often use the same word to mean different things, or conflate key concepts. My colleague Lauren once said, memorably, "I wish I could make a new word for 'brand.'"

To ensure we have a shared vocabulary, here's how we'll define these terms in this book.

- **Brand:** Your customers' perception of and relationship with your company, products, and services.
- **Branding:** A disciplined process of creating a differentiated identity in customers' minds.
- **Brand identity:** The tangible and intangible elements (logo, visual system, UI/UX, tone and voice) that shape how your product is recognized and experienced.
- **Brand campaign:** A specific marketing activity designed to support overall business goals.
- **Brand building:** Activities across all touchpoints—from product experience to marketing campaigns—that establish and strengthen customer relationships.

As Jeff Bezos put it, "Brands for companies are like reputations for people and reputations are hard earned and easily lost . . . you make a promise and then fulfill the promise."[10]

Steve Jobs also recognized that the most powerful brands are about earning customer trust. "To me, a brand is one simple thing, and that is trust. Your customers trust you. And so brands are like bank accounts. You can have withdrawals and you can have deposits. So if a customer has a great experience, they buy an iPod and they love it. That's a deposit into our brand account."[11]

I believe building a brand is fundamentally about forming meaningful relationships with customers through your products and services. Your brand is not about what you say about your company, but what customers say about the value you deliver to them. Your brand exists in the minds of customers whether you actively shape it or not. It's the sum of their perceptions, experiences, and relationships with your products or services. This is what brand power is about.

This definition reveals a simple truth: You cannot not communicate your brand. If you don't communicate intentionally, customers will form their own perceptions from scattered, disconnected interactions. The choice isn't whether to have a brand; it's whether you actively design how customers perceive and relate to your products, or leave it to chance.

Brand and Product Are Inseparable

Your brand is the heart of your products. It's the "why" behind what you build. It's not only inseparable from your product but also elevates it.

For example, Ring integrates brand efforts with product development and offerings to shape customer experiences. Early brand efforts gave the doorbell innovation meaning and differentiation. That made the difference between being rejected on *Shark Tank* and becoming one of the world's biggest home security brands. The early brand work wasn't just a creative exercise. The brand's narrative informed the product roadmap and provided a sharper focus that ultimately led to long-term business success.

That's a fundamental piece that people often misunderstand. I remember vividly an important brand architecture review with Amazon's senior leadership. The meeting didn't go as well as planned because the entire conversation mostly focused on logos and visual identities, whereas the product roadmap and business growth were not properly incorporated to drive toward a decision and alignment on the final recommendation. This experience reminds me of a deeper issue: Many teams still view brands and products separately. But to shape what customers see and feel, brand strategy must be fully integrated with product and business goals from the start.

This challenge is particularly unique to tech products and services. Traditional consumer goods have long relied on packaging and advertising to drive differentiation and build customer relationships. For instance, Coca-Cola has been around since 1886, but its product formula hasn't changed much. Coke's innovation may have to rely on its mass marketing and advertising machine to refresh its brand image and remind customers that Coke is always there to open happiness at moments in life when you need it. The reality was that there weren't significant differences in their products. Customers often couldn't taste the differences between Coke and Pepsi, so the brand becomes the deciding factor for purchase decisions.

But tech products are fundamentally different from traditional consumer-packaged goods. Consider smartphones: iPhones and Android phones have real functional differences—the operating system, design, ecosystem integration, and so on. Even among brands that are based on the Android system—such as Samsung, Xiaomi, Google Pixel, Motorola, and OnePlus—all differ in their actual product features, price points, and user experiences. So, when you market the product, you naturally market the brand.

In tech, the product experience is the brand experience. Every touchpoint, from product naming, packaging, and website, to acquisition and onboarding, impacts customer perceptions and their choice between you and your competitors. All these touchpoints are moments where you can either win over customers or lose them. This is why you need to broaden your understanding of "product" and think about the entire customer experience as your product.

Building a great product experience is brand building. You can't treat brand as a separate layer added after product development; it's fundamental to building strong customer relationships and driving value.

When you separate brand and product, you risk creating generic products that lack meaningful differentiation and struggle to gain traction.

Myth #2: "Product Innovation Alone Will Drive Success."

Many product builders assume great technology alone can scale a business. While innovative tech may attract early adopters, it often fails to engage the broader market to drive sustainable growth. Why? Because growth requires more than innovation—it demands education, connection, and trust to drive adoption.

In the tech world, startups often originate from innovative and disruptive technologies by engineers or technologists. My experience with Amazon Key is a typical example of how new tech products and services begin with exciting innovations but face hurdles as they try to launch and scale.

Almost all innovations face three critical challenges:

1. Finding early customers and achieving product-market fit
2. Expanding beyond early tech adopters to mainstream markets
3. Building a sustainable, enduring business

Brand efforts can directly address each of these hurdles. This is why early-stage ventures must prioritize brand building to navigate these challenges at different stages. The power of brand building is the difference between innovation with potential and realized success, as we see from Ring's founding story.

The Leap from Tech to Meaningful Products

It's natural for companies to focus on building interesting technology first, but there's a difference between innovative technology and a fully realized product. A great product needs to solve a customer problem and add value to people's lives, and brand efforts can help turn innovative technology into meaningful products. Positioning, branding, storytelling, and better experiences can shape the product development process and make your product more appealing to your target customers.

This sentiment is echoed by Elliot Cohen, cofounder of PillPack, who shared his perspective during a conversation we had in late 2024. Cohen emphasized that founders should think about brand from the very beginning. In his view, it's essential to get the product story right on day one, as the story itself can influence how you build the product, the features you create, and the types of customers you solve problems for. But many companies struggle with storytelling.

Usually founders have a vision for the "why," but the big challenge lies in how to articulate that vision and make it actionable. It takes creativity and imagination to translate technology and product ideas into a tangible story that deeply connects with customers. For example, Ring's narrative isn't merely about a doorbell; it's about connecting people with what's important to them: their homes and communities. "Always Home" speaks to the human and community connection and makes Ring unique and different. By elevating the doorbell beyond its functional use, the product fosters the emotional connection.

A great product always starts with a meaningful customer problem. It can't just be a technology solution—it must deliver unique value that connects with people on both emotional and functional levels.

Today, we are seeing a similar challenge with AI. As we witness an explosion of AI-enabled products and services—from OpenAI's ChatGPT to Anthropic's Claude, Apple Intelligence, Meta AI, and DeepSeek—the challenge isn't just building know-it-all models, but also solving specific customer problems and delivering meaningful value. Many AI products and initiatives are led by scientists and engineers, and this often results in technical jargon and undifferentiated product positioning. Terms like "language model," "generative AI," "deep learning," and "neural network" can be intimidating to everyday users. Bridging this gap requires translating innovation into differentiated, relatable customer value.

Without a clear brand and product strategy, these new AI products risk being seen as just another AI tool. Brand-building efforts can establish unique value propositions, help people understand specific benefits and use cases, and move beyond generic AI-powered narratives. As we see from Ring's transformation, when you invest in establishing distinctive positioning and storytelling, you can translate technical features into meaningful products that resonate with people.

The Leap to Cross the Chasm

Even the most revolutionary products require effective marketing and brand building to achieve mass adoption and success. This is because, as Geoffrey A. Moore explains in *Crossing the Chasm*,[12] every innovation has an adoption cycle with different customer segments: early adopters, early majority, late majority, and laggards. Some people will naturally adopt your innovation earlier than others. To cross the chasm between these customer segments, you need innovative product and marketing strategies at different stages.

While product innovation is imperative, it's not enough to drive mass adoption. A brand acts as the bridge, translating complex technological advancements into compelling value that mainstream customers understand and want.

I experienced this firsthand when I worked on Apple's global marketing and advertising. These days, all sorts of people use Mac computers everywhere, but that wasn't the case 20 years ago. Back then, the average consumer believed that Mac was the right computer only for creative minds, not for them. At the time, Mac had less than 5% market share, whereas most computer customers bought a desktop or laptop PC.

To change preconceived notions, Apple and its advertising partner TBWA\Media Arts Lab developed the "Mac vs. PC" campaigns, which ran from 2006 to 2009. Do you remember the stuffy PC character played by John Hodgman and the cool Mac played by Justin Long? Along with educating everyday users about the value of Mac, this campaign helped reinvent the Apple brand. As a result of this simple, entertaining, and memorable campaign, Mac's market share jumped, paving the way for Apple's continued growth as a brand for the masses, not just elite users.

You see, even for an innovative, well-designed product like Mac, customers still needed to understand the benefits before switching from PC. If they hadn't run this popular campaign for four years, Apple might not have been able to grow beyond a niche product to gain significant market share.

The Mac's journey proves that even revolutionary products need strategic brand building to cross the chasm and reach mass audiences. Without it, innovation can't achieve its full potential and drive scalable growth.

Myth #3: "We Can Build a Brand Through Advertising."

In the past, companies could build strong brands primarily through advertising campaigns. This worked for traditional consumer goods where products were commodities, and advertising created the differentiation. In 2006, when I began my career as a brand strategist at Wieden+Kennedy (Nike's global creative partner), advertising was still the primary tool for building a strong brand. At that time, digital was emerging, but television was still the most effective medium for delivering messages and connecting with customers. Brands like Coca-Cola and PepsiCo were some of the largest advertisers, and they relied on advertising to infuse their products with meaning and differentiation.

But this approach doesn't work for tech products, where the product experience itself must be the source of differentiation. Tech products require elevated product marketing that is consistent with product experience. Advertising campaigns can amplify the inherent product truths, but you cannot *create* differentiation through advertising. These challenges are compounded by the reality that advertising is increasingly oversaturated, making it harder for any new products to break through.

Advertising Becomes Saturated

Today, increased advertising spending no longer translates into customer loyalty. Consumers are bombarded with ads and feel increasingly overwhelmed. And performance marketing channels don't support the creation of sustainable demand for the future.

Every year, investment in advertising has continued to increase, leading to market saturation and intense competition for customer attention. US advertising investment increased by 43%, from $294 billion in 2020 to an estimated $421 billion in 2024.[13] Globally, advertising spending was forecast to reach $754.4 billion,[14] growing by 5% in 2024, compared to 3.3% in 2023. However, this increased investment doesn't necessarily translate to increased effectiveness, as consumers have become more and more adept at tuning out advertising messages.

According to System1's Ad Ratings, 48%[15] of all advertising is now dull and ineffective. Among the more than 100,000 ads tested in the study, almost half didn't evoke any emotions. Consumers are bombarded with ads and feel increasingly overwhelmed. As of early 2023, surveys revealed that 31% of US adults used an ad blocker to avoid ads and protect their privacy,[16] making it harder for advertisers to reach consumers.

Digital Platforms as New Middlemen

On the other hand, advertising has become more measurable, inherently digital, and performance driven. With the convenience of buying digital and social ads, there has been an overt emphasis on lower-funnel performance advertising (the kind that drives clicks) over general awareness and perception advertising (the kind that changes perceptions).

Companies now allocate most marketing investments to paid search advertising to buy traffic. For instance, in 2023, Google Search, Meta (Facebook and Instagram), and Amazon Ads captured 60%[17] of the $680 billion global ad spend, showcasing the scale of investment in performance channels. In the travel category alone, Airbnb, Expedia Group, and Booking Holdings each spent billions of marketing dollars to compete for traffic.

This ad model is like "digital rent"—a cost of selling through the platform rather than a means of generating future demand, prioritizing short-term sales over winning sustainable customers. As these platforms increasingly act as the new middlemen, brands and products compete for limited digital shelf space to get in front of potential customers.

As Colleen Decourcy, chief creative officer of Snapchat, puts it, "Without a brand, and a sticky message, or a point of view people care about with their hearts and minds, the second your dollar stops hitting that tray, your performance marketing stops working."[18]

This reality is forcing companies to rely less on advertising to push the product out and compete in downstream performance advertising spaces. Instead, they have to find more innovative ways to connect with customers. This means companies have to prioritize product differentiation and value and build better experiences—in other words, building their brands.

For example, Ring's success wasn't built through traditional advertising. Ring turned its product into a powerful marketing tool. Its video sharing feature and Neighbors app became an organic content marketing machine, amplifying user generated content (UGC) and creating significant word of mouth that accelerates Ring's adoption in local communities.

To escape this "digital rent" competition, companies must innovate and look at growth opportunities more holistically across acquisition, engagement, and retention to strengthen customer relationships beyond paid advertising.

Why We Need a New Approach

The three myths we've explored all stem from outdated thinking about how brands are built. The goal of building a brand has always been about winning customers, but today's tech landscape requires a fundamentally different approach.

Transformative shifts in consumer behavior, distribution channels, media consumption, and business models have reshaped how new products launch, grow, and connect with customers. The new reality demands better integration to capture the opportunities in front of us. In a world where most tech companies rush to chase AI, efficiency, optimization, and quick wins, it's easy to lose sight of customer relationships that drive long-term brand power.

Companies like Ring succeed because they defined and integrated brand strategy from day one, embedding it into every stage of a new product's life cycle. This approach created a built-in competitive advantage that informed product decisions, customer experience, and go-to-market strategy, enabling the company to scale to become a category leader.

From early startups to the most successful tech companies, we've seen that building a strong brand has been proven to increase differentiation, accelerate growth, and generate lasting value. From 2006 to 2024, the value of the world's top 100 brands increased 474%, and the threshold to become one of the top 100 brands increased 354% from $4 billion to $19 billion. Brand building for tech products is particularly impactful. On the world's top 100 most valuable brands,[19] tech companies dominate the list, contributing $1.2 trillion of the top 100's $1.4 trillion growth[20] in 2024.

The good news is that this type of brand power is accessible to companies at every stage. A startup needs to build a new brand to focus on driving differentiation to break into an established category and attract new customers. A mature business may need to strengthen an existing established brand to drive repeat purchases and long-term growth. When you build a strong brand, you win customers' hearts and minds—you drive both product and business success. This isn't just a goal for product builders, marketers, and business leaders—it's the ultimate goal of any business.

To succeed in this new environment, you need a new playbook—one that transforms how products and brands are developed and how businesses create value. This isn't just a process change; it's a fundamental strategic shift. We'll discuss how you can drive this shift in the next chapter.

--------------------------------- **Chapter Summary** ---------------------------------

Brand is not fluffy but a fundamental driver of business success that must be integrated from day one. Building a brand is ultimately about winning customers. This isn't just a goal for product builders, marketers, and business leaders—it's the ultimate goal of any business.

The tech world is in the grip of three pervasive myths that explain why many companies undervalue brand building:

1. "Brand is separate from product."
2. "Product innovation alone will drive success."
3. "We can build a brand through advertising."

These myths hold back even the most successful tech companies, shaping how we think about and invest in brand building and product development.

Brand and product are inseparable. In tech, the product experience is the brand experience. A brand is not a superficial layer added after product development. Building a great product experience is brand building.

Product innovation alone is not enough to transform new products into scalable businesses. Only when paired with effective marketing and brand building can innovation bridge the gap between theoretical potential and realized potential, driving consumer adoption and long-term success.

Companies can no longer rely on post-launch advertising to build lasting relationships with customers. Advertising has become oversaturated. Digital platforms increasingly act as middlemen. Companies must find new ways to connect with customers outside advertising channels.

We need a new approach to brand building. Consumer behavior, distribution channels, media consumption, and business models have reshaped how new products launch, grow, and connect with customers. This requires a strategic shift in how businesses create value.

Chapter 2

Building In Brand Power from the Start

*From Brand as a Nice-to-Have
to Brand as the Heart of Your Products*

Most companies treat brand as a nice-to-have that can wait. But winning from day one is about not losing from day one. *Brand Power Built In* flips this assumption by integrating brand strategy directly into product development. This chapter introduces the framework, its three core principles, and your roadmap for implementation.

The Trade-Offs You Don't See Until It's Too Late

Every founder starts with a vision. They imagine a better way of doing things, a groundbreaking product, or a solution that will change lives. But somewhere between the vision and execution, it's easy to get lost in the daily pressures and trade-offs.

I recently spoke with a founder about an AI-enabled app he had been building. He and his team spent more than two years creating a new product that was supposed to change the entire category. He showed me his early beta version, and after testing it out, I wasn't sure what it would do for me as a customer. I didn't understand the purpose. I gave him my honest assessment and feedback.

As a founder who deeply values brand thinking and product design, he was surprised, almost shocked, that his vision didn't show up in the app experience. He captured this challenge candidly in an email to me as we discussed founders' daily struggles:

> I asked myself, "Why is this the case? Why is it missing these very obvious things?" I think the answer lies somewhere in the fact that when you try to build something, there are so many difficult and time-consuming tasks that you have to make trade-offs. Sometimes, things that feel ethereal or less tangible, like your story and your reason for being, don't make it through. Those are the harder things to communicate, and those are the more difficult things to have other people understand. It's very easy for someone to say, "We need to build a website and it needs to do these things," but it's hard to say, "It has to have the story and it has to have this feeling so our customers will feel a certain way."
>
> I've had to trade them for very short-term progress. I can now see how it's a slippery slope—that you can wake up one day and realize you just didn't get what you meant to get.

This raw perspective captures a challenge I've seen time and time again: When startups move fast and prioritize short-term execution, the brand they intended to build never materializes. Even well-intentioned founders who didn't ignore brand at the start often find that they've somehow lost it in the process.

This founder's experience reveals something deeper than execution challenges. Every day that you delay brand efforts, you're not just deferring decisions, you're making irreversible ones. Even established tech companies struggle to maintain their brand focus amid pressure to ship new features and drive growth.

Winning from day one is about not losing from day one. It's easy to lose customers and difficult to win them back. The difference between market leaders and failed products often comes down to seemingly small choices made early in development. Unclear positioning, a poor product name, bad packaging, generic value proposition, a low-quality website, inadequate onboarding experience . . . all of these seemingly "small things" add up and impact how customers see you and their choices between your product and your competitors' products. These are the critical moments where customer perceptions and relationships are formed, where brand power is built, and where winning and losing are defined.

It's usually easy to calculate wasted spending, but this is the type of intangible "wasted opportunity" and "lost value" that people don't see. When you miss these moments, you lose customers easily. That's a hit to your financial value, your market share, your profitable margin. That's the hidden value most companies don't see until it's too late.

The opportunity isn't waiting for the right moment or securing more resources to build your brand power—it's about building it into everything you do from day one. This is exactly what *Brand Power Built In* was designed to do.

The Idea: *Brand Power Built In*

Bringing brand thinking to product is not natural for most organizations. In many cases, brand thinking is either missing or remains superficial. The traditional approach is linear: "Here's the product—now go sell it." Marketing is often seen as a sales and service organization, not part of the building phase. As a result, marketing is usually brought into the process when the product is almost ready to ship, thereby creating silos and undermining the potential to make an impact earlier.

The *Brand Power Built In* approach flips this process: Embed brand thinking and marketing early into the product. Instead of treating brand building as a downstream marketing activity, you begin with the end in mind—*building customer connections early on, using the same vision to guide both product development and brand development.* This approach transforms the product development process, enabling teams to create differentiated offerings that deeply connect with customers and make them easier to go to market. This approach creates better products that attract customers naturally, rather than waiting to push products out to people, often relying on expensive marketing and advertising investment to drive demand after launch.

OLD	NEW
Build it $\rightarrow$ Market it	Build it + Market it $\rightarrow$ Market it

This approach is rooted in three core principles:

Principle #1: Product experience is the most powerful brand
  builder
Principle #2: Your product is the entire customer journey
Principle #3: Brand building is a shared responsibility

It's about embedding customer connection and emotional resonance
into product development and experience design from the start, not after
it's done.

It's about integrating long-term brand-building efforts with the iterative product development process, not treating it as a separate activity.

It's about investing in meaningful customer experiences and relationships across the entire journey and product life cycle, not just one
part of the journey.

It's about leveraging what you're already doing and making everyone's collective efforts more efficient, not about investing more budget
on brand.

This approach fundamentally changes how products and services
are developed, launched, and scaled. It requires a mental shift to bring
together the entire organization to create the foundation to make the
leap from good to great. But we know that this mental shift has worked
for some of today's most successful companies.

That's how Ring changed the home security industry, how Airbnb
revolutionized the way we travel, how Rivian broke through a crowded
car market, how PillPack reimagined how medications are managed and
delivered to patients, how Notion redefined workspace collaboration,
how Robinhood democratized finance for everyday investors, and how
Figma made design accessible to everyone. We will explore all of them
in the chapters ahead.

If you want to disrupt an existing industry, you must be innovative in your product, brand, and the entire customer experience. At its core, brand thinking is the strategic process of defining what you stand for and bringing it to life through your product and experiences that create deeper customer relationships. It starts with specific customers in mind and a clear vision for the functional and emotional value you want to deliver.

This proactive brand thinking enhances the product experience, makes it easier to sell when you go to market, and ensures business longevity and durability.

Let's unpack how each principle works in practice.

Product Experience Is the Most Powerful Brand Builder

The most powerful brand experiences occur outside of advertising channels—through product experience and ongoing engagement. These underinvested areas are where companies have the most ownership and leverage: brand development, product experience, UX design, and customer relationship building.

This is because your brand infuses meaning, purpose, and life into every aspect of your product experience. It's your product's DNA and the engine that drives everything. This is why you can't finish the product and then put the heart back. It needs to be embedded early on when the product is still in development.

As Sir John Hegarty, an advertising legend, states, "What's more important for your brand? Your product or your marketing? The greatest companies all feature a product-centric mindset . . . Never ignore product. It's the only part of your business that customers care about."[1] Hegarty's point reveals the fundamental truth that your product experience is your most honest form of marketing.

How Product Drives Brand Value

A company's product often provides the most frequent touchpoint with customers. Think about the brands and products you can't live without in life: your phone, your music app, your payment methods, your workspace tools, your streaming services, and so on. Your daily interactions with these products and services directly influence your perceptions and purchase decisions. Seventy-three percent of customers cite customer experience as a major factor in driving purchasing decisions. Sixty-five percent of customers believe that a positive customer experience is more influential than great advertising.[2]

For example, Amazon's approach to brand building is all about investing in customer experience and relationships. Neil Lindsay, the former VP of Worldwide Prime and Marketing at Amazon, wrote, "I also think our customer relationship is that of a 'small b' brand. The relationship they have with our brand is an output of the customer experience. A 'capital B' Brand might have a manifesto that defines the products they build and so forth. But for us we're 'capital C'—Customer obsessed, which puts the focus on the things we need to invent."[3] Amazon shows us how a successful tech brand is built differently from the past, mostly in the customer experience, reinforcing the need to broaden our understanding of brand and product efforts.

Tony Fadell, the founder of Nest, understood these engagement opportunities when he and his team built the Nest Thermostat. In the beginning, the entire team primarily focused on building the hardware—the cool thermostat gadget—and didn't put much effort into building the Nest app. But when Tony considered the entire customer journey, he realized that 10% of the customer experience came from marketing, PR, and advertising, 10% from product installation, 10% from interacting with the device, and 70% from the Nest app on people's phones. This deep consumer insight led his team to build a new Nest app to work in tandem with the thermostat, create the user manual in the packaging, and develop marketing and retail promotion to tell customers about it.[4] If the team had ignored or downplayed the Nest app to focus only on the hardware, they would have neglected 70% of the entire customer experience and an important opportunity to build the Nest brand.

From Amazon to Nest, we see how product directly drives brand value. When you offer the best products and customer experiences, people will talk about them, write online reviews, and share them with their family and friends. Word of mouth is powerful to drive advocacy and create momentum, ultimately helping build more resilient businesses.

UX Becomes the New Face of the Brand

In the past, branding was mostly manifested in traditional advertising, packaging, retail stores, and physical displays. Today, user experience has become the primary face of any tech brand. As we increasingly interact with digital products and services, branding in UX plays a more critical role in conveying brand values and promises.

I witnessed this transformation firsthand through my work with Apple. When the iOS App Store launched in 2008, only 500 apps were available to iPhone users. Today, there are over 4 million apps available across iOS and Android platforms. Americans now spend an average of 5 hours and 24 minutes on mobile devices every day. Our phones have become the first thing we touch in the morning and the last thing we use before we go to bed. This shift marked a pivotal moment in brand building, when the focus began transitioning from traditional advertising to product and user experiences.

The more time people spend on the internet and personal tech devices, the more critical digital and mobile platforms become as brand touchpoints. The user experience now represents an underinvested area with immense potential to deepen customer relationships. For example, if you're developing a new app, your app will sit among millions of others on the digital shelf of the Apple and Android app stores, where users scroll quickly. If the name or icon doesn't connect, people may never even open it and experience the product, proving that branding and UX begin before the first tap.

But how can you stand out in highly saturated internet and mobile environments where you only have 50 milliseconds[5] to make a great first impression? In that very brief moment, your brand identity, messaging, and user experience must work together to immediately convey value. This requires a strong point of view, distinctive identity, and cohesive storytelling across all touchpoints.

Bringing Brand Thinking to Experience Design

The opportunity is to bring brand thinking to product and customer experience design from the very beginning. Building brand power can elevate a product to become more distinctive and memorable. Without this differentiation, products risk being generic or the same in a crowded market where customers already have lots of choices.

Consider businesses like Craigslist or Booking.com. They solve customer problems effectively, but their relationships with customers remain transactional. In contrast, brands like Airbnb and Apple go beyond utility. These products are not merely functional, but also memorable and meaningful, creating emotional connections with customers.

Digital products must first offer utility value. They have to work with reliable and consistent performance. But at the same time, they need to make customers feel something. Balancing these dual requirements of utilitarian function and emotional connection is challenging to achieve, but exciting when it works.

For example, when people use Slack, the brand also lives in user flows and daily interactions. Slack's playful messages and custom emojis aren't just features; they are brand moments that happen hundreds of times per day, directly shaping how customers perceive the product.

When you infuse brand power into a product, you transform it into a differentiated offering with personality, purpose, and emotional resonance that motivates prospects to choose you over the competition. This requires you to bring positioning, branding, and storytelling that usually inform brand development and campaigns much earlier to create deeper engagement and connection through the product and user experience. We'll see how to bring this to life in chapter 5.

Your Product Is the Entire Customer Journey

The *Brand Power Built In* approach integrates brand thinking into both the core product offering and the end-to-end customer journey. Your product isn't limited to the core offering but includes every customer touchpoint. It's no longer just about upper-funnel awareness or lower-funnel acquisition. It's about shaping perceptions, behaviors, and relationships across acquisition, engagement, and retention. The entire customer experience, start to finish, becomes part of your product experience.

Business leaders often debate about this question: Between customer acquisition, product engagement, and customer retention, which part of the customer journey shall we invest in, and how can we get the most ROI? Since companies rarely have the resources and budget to do everything at once, a holistic approach that considers the entire customer journey is essential for making investment decisions. This integrated approach helps you understand where your customers spend the most time and engage most frequently, so you can identify needle-moving initiatives among other priorities.

At its core, successful brand building is intrinsically linked to a customer-centric approach. Mapping the customer journey gives you a laser focus on the customer and their needs, which helps you identify opportunities and gaps in the experiences you want to create and the connections you want to form.

Bringing Cohesion to Fragmented Experiences

Today, as consumer touchpoints grow more fragmented, you need to focus on the aspects of the journey you can control and design cohesive experiences. Why? Every touchpoint matters. When you have fragmented experiences, you risk losing customers. That leads to lost financial value, less market share, smaller profitable margin . . . That's the hidden value most companies don't see.

The big challenge is that, in most organizations, no one oversees and manages the end-to-end customer experience. Product development, UX design, and brand development operate independently. When they don't collectively create the end-to-end customer experience, it's difficult to build unified experiences that truly resonate with customers.

Let's use email communication with prospects and customers as an example. In many companies, product and operation teams may own some parts of the customer communications such as ordering, returning, and welcome emails. The marketing team may own communications dedicated to acquisition and retention, such as new product announcements and win back campaigns. Without coordination and centralized oversight of execution, customers may receive excessive or conflicting emails, leading to confusion and a potential loss of trust. That's an example of how you potentially lose customers. The next time you're about to send out yet another email, remember that there's probably a frustrated customer on the other side. How many times is your company emailing them *in total*?

The opportunity lies in helping companies understand the strategic importance of improving customer experiences and designing a process to unite the efforts of various departments. Brand thinking brings clarity and cohesion to increasingly fragmented channels and environments through a strong brand identity, cohesive customer experiences, and consistently reinforcing your value.

Brand Building Is a Shared Responsibility

In other words, building a long-term brand is not just a marketing activity but a shared responsibility across departments. This demands a cultural shift in how teams innovate and collaborate to design and deliver great customer experiences. This doesn't mean accountability is diffused, but rather that every team must understand how their work directly impacts customer perceptions and behaviors.

Breaking Down Walls

The traditional organizational structure and process design hold many companies back. The brand team often sits within the marketing department, but product development, UX design, marketing, customer service, PR, and anyone else whose work plays a role in shaping consumer perceptions and influencing behaviors all contribute to building a strong brand. This is a legacy of the old days, when marketing was the sole function driving brand activation.

In consumer packaged goods (CPG) companies like P&G, brand managers typically oversee mass advertising, brand stewardship, product sales, and their respective profit and losses. But in the tech industry, the product team plays a crucial role in shaping customer perceptions through product development and user experience, though many teams haven't fully realized the potential impact on the brand.

If you're a product manager or UX designer, you may already have shaped your brand without even realizing it. Because you don't typically see yourself as a brand builder, you often focus on delivering product features, creating UX wireframes, and working with engineers to meet technical requirements.

Similarly, we don't often associate engineers with marketing and brand, but their work in building and delivering the customer experience directly reinforces the customer's perception of value and trust.

Engineers shape the brand through less familiar touchpoints like app performance, software functionality, and tech-driven services such as notifications. These areas are traditionally not seen as brand work but critical to customer perceptions and behaviors.

Uniting All Functions Around the Customer

From design to UX, from in-product copywriting to marketing copy, a cohesive customer experience is crucial for building stronger products and more successful businesses. This is not about investing more budget to build the brand. It's about leveraging what you're already doing and streamlining processes to reduce fragmentation, be more efficient with resources and investments, and be more customer focused.

This requires a mindset shift in how teams collaborate across functions. Marketing leaders must be integrated into the early product development process, while product leaders must understand the power of brand as a strategic connection enabler. Leadership must establish a customer-centric culture that fosters this cross-functional collaboration across the end-to-end customer experience. Then, when new products go to market with this integrated approach, they'll be more likely to connect with potential customers and stand out in the market.

No one sees this cultural shift better than Dylan Field, CEO of Figma, whose design tools have fostered collaboration in digital spaces for successful companies such as Stripe, Netflix, and Spotify. Field has seen firsthand how working together in digital spaces through a shared tool moved people from a mindset of "my ideas" to "our ideas." In his words:

Unlike physical spaces, digital spaces have no walls: by default they are non-hierarchical. Everyone is invited to brainstorm, build, and play together . . . At its best, Figma is much more than a digital extension of our physical self—it's an invitation to leave ego at the door and create shared consciousness with others. It's an opportunity to embrace the messy bits of creativity, to celebrate failure and bring more people into the process along the way.[6]

This insight highlights the cultural shift needed to break down walls and foster collaboration to create better products and experiences. It embodies the transformation required to build truly customer-centric organizations.

The Six Building Blocks for Implementation

Building great products and brands is an inseparable journey. But where do you start? How do you implement this collaboration?

The next six chapters will provide practical tools to integrate brand strategies into every critical stage. This journey mirrors the life cycle of a new product—from development to launch to scaling—while showing you how to bring product and brand together and embed customer connection throughout the process.

Build

1. **Develop Your Positioning Strategy**
   An effective positioning strategy is your North Star for product and brand development. With a clear vision, a defined market position, and a customer problem to solve, you can build an intentional business from the outset.

2. **Establish Your Brand Identity**
   Brand identity is a strategic lever for business value creation. From crafting your story to developing a distinct naming

The Six Building Blocks

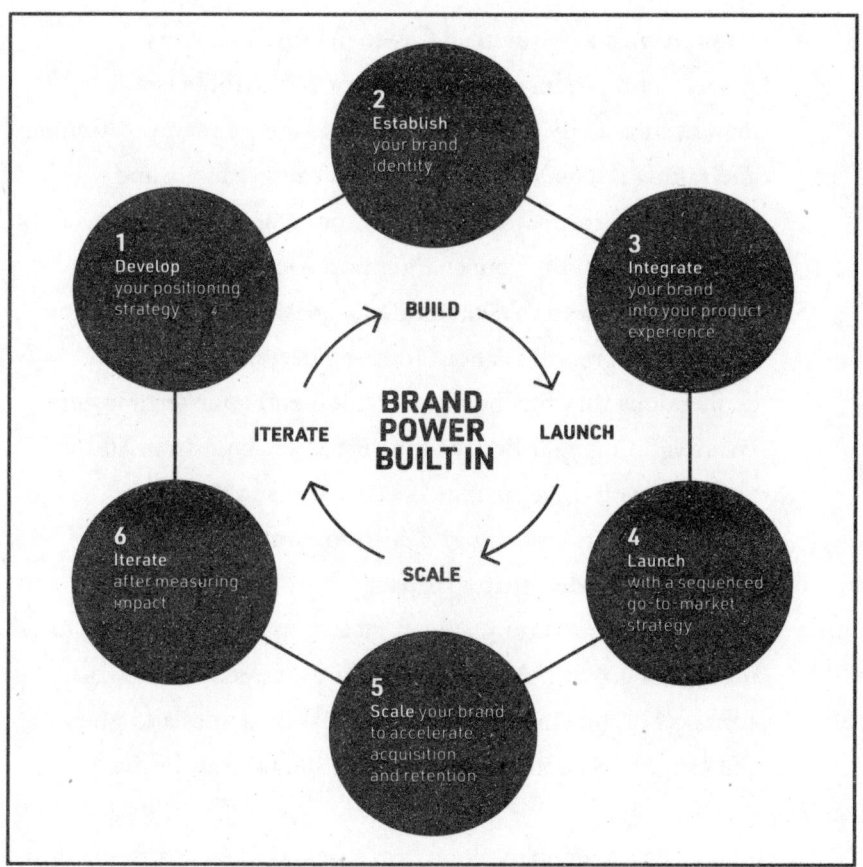

and visual identity, branding gives your product meaning, differentiation, and intrinsic value.

3. **Integrate Your Brand into Your Product Experience**
   Your product experience is your most powerful brand-building tool. When integrating brand foundations into the product experience, you can transform fragmented customer experiences into more differentiated products with personality and purpose.

Launch, Scale, and Iterate

4. **Launch with a Sequenced Go-to-Market Strategy**
   A successful product launch is about establishing product-market fit as early as possible. By leveraging brand positioning and taking a sequenced strategy, you can gradually and intentionally form stronger connections with your target audience to solidify your market position.

5. **Scale Your Brand to Accelerate Acquisition and Retention**
   Sustainable growth is about forming deeper customer connections that fuel both acquisition and long-term loyalty. By integrating your brand with customer acquisition and retention, you drive sustainable demand and transform transactions into meaningful customer relationships.

6. **Iterate After Measuring Impact**
   Brand power requires nuanced measurements beyond the usual metrics of growth. An integrated measurement framework connects brand strength with financial outcomes and gives you the insights needed to optimize and iterate your business.

Brand building is an ongoing process that requires consistent efforts. There's no magic formula. There are no shortcuts. The six building blocks are foundational tools to help you move toward the best customer experiences and build a more valuable business that will stand the test of time.

———————————— **Chapter Summary** ————————————

Brand Power Built In **is rooted in three core principles:**

1. Product experience is the most powerful brand builder.
2. Your product is the entire customer journey.
3. Brand building is a shared responsibility.

Product experience is the most powerful brand builder. In the past, branding was mostly manifested in traditional advertising, packaging, retail stores, and physical environments. Today, UX has become the new face of the brand, representing an underinvested area with the biggest potential for building customer relationships.

Your product is the entire customer journey. Your product isn't limited to its core offering; it also includes every customer touchpoint—every detail that may shape perceptions, behaviors, and relationships across acquisition, engagement, and retention. The entire customer experience needs to be seen as your product.

Brand building is a shared responsibility. This mind shift often requires a fundamental change in how teams collaborate across an organization. This integrated thinking can empower the entire organization to be more customer-centric and create better products and experiences.

The six building blocks provide a roadmap to integrate brand strategies into critical product stages. This journey mirrors the life cycle of a new product—from development to launch to scaling—while showing how to bring product and brand together and embed customer connection throughout.

PART II

Build

How do you build brand power and customer connection into product development? Where do you start? The answer lies in bringing brand strategy upstream and adopting a "selling" mindset—focusing on what makes your product unique and lovable to customers—rather than waiting until the go-to-market phase. This is a strategic opportunity most companies overlook.

We'll start the journey with the first three building blocks to build a better brand and product experience together:

1. Develop your positioning strategy.
2. Establish your brand identity.
3. Integrate your brand into your product experience.

By embracing the first three building blocks in the building stage, you'll transform the development process, creating differentiated products with personality and purpose while laying the foundation for go-to-market success.

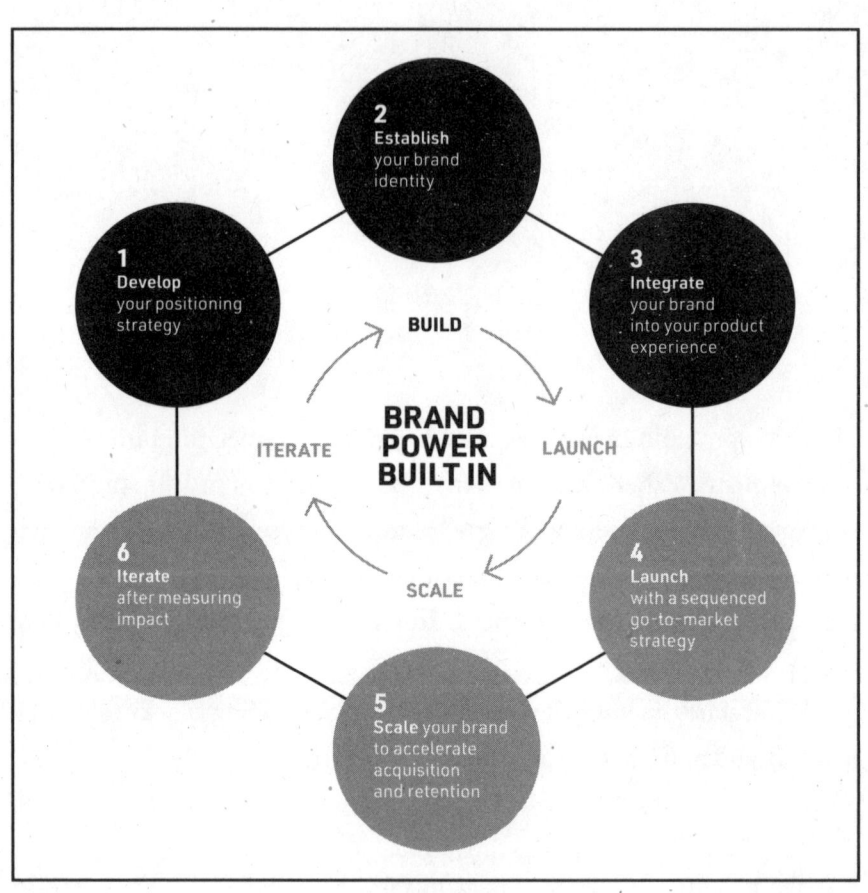

Chapter 3

Develop Your Positioning Strategy

From a Marketing Tool to a Foundational North Star

Why do so many great products fail to establish their position in the market? Companies often struggle to connect their innovations with customer problems and unmet needs.

At the heart of the *Brand Power Built In* approach is your positioning strategy, which serves as your North Star to guide product development, business creation, and brand development from the very beginning. This chapter shares strategies to craft a positioning framework that helps you drive meaningful innovation and avoid common pitfalls.

The Rivian Adventure to Become the Most Loved Car Brand

How did an unknown car startup become the most loved automobile brand in America? In 2024, Rivian surpassed Tesla in brand loyalty and became the No. 1[1] most loved car brand in the United States, not only within the EV category but across the entire auto market. The company achieved this recognition only three years after the first model went to market in 2021. But Rivian's journey wasn't easy, and it took the company almost nine years to launch its first product.

RJ Scaringe, an MIT-educated mechanical engineer, founded Rivian in 2009—the same year GM and Chrysler filed for bankruptcy. Before the official debut at the Los Angeles Auto Show in 2018, no one had heard of Rivian, and the EV market looked very different from what we see today. Tesla dominated the market, while traditional automakers were only beginning to roll out their own models. If customers wanted to buy an electric car, they didn't have many choices then.

"We're building this company because it doesn't exist."[2] That was on the back of the wall when RJ started the company in a warehouse. The car industry is extremely complex and building everything from the ground up is almost unimaginable, but RJ wasn't constrained by dogmas. He founded Rivian to redefine expectations through the application of technology and innovation while transitioning the world toward sustainable energy.

In 2018, the company unveiled the first fully electric pickup truck (the R1T) and an all-electric seven-seat SUV (the R1S), and then brought them to market in 2021. The R1T model started at $61,500 after federal tax credit, and the R1S SUV began at $72,500.

Rivian broke through a crowded car market by identifying an underserved customer segment in the category: environmentally conscious outdoor adventurers who desire pickup trucks and SUVs and are also willing to pay for luxury features of premium vehicles. This brand position was the foundation of Rivian's product and go-to-market strategy. At the time, no major automakers were addressing the intersection of electric, adventure, and luxury. Rivian strategically identified this gap and turned adventure into an electric opportunity.

Rivian tapped into a universal human desire—an adventurous spirit—and has since built a brand with a mission to keep the world adventurous forever. As RJ explained, "A lot of the aspirational vehicles tend to be more like an Armani suit. These brands do have things that are more functional, almost like an Armani leisure wear, but there's no one that has built up and focused on that Patagonia-like brand position."[3]

Being electric alone is no longer a sufficient differentiation point. As a technology-driven car brand, Rivian builds vehicles with adventurers at the core of design and engineering decisions. The focus on user experience with software-centric design capabilities is a key differentiator. Its early focus on adventure and luxury pickups and SUVs paid off. In 2023, Rivian R1T earned the J.D. Power Award[4] for the highest satisfaction and best ownership experience among premium electric vehicles. This achievement wasn't possible without a clear vision, an unwavering focus on customer experience, and millions of decisions made by thousands of engineers and designers working behind the scenes to turn that vision into reality.

The company's focus on experience extends to its charging stations. For example, in its first-of-its-kind charging outpost near Yosemite National Park, customers can truly relax and refresh, making charging less like a chore and more like a part of the adventure.

Looking back, Rivian carved out a segment of customers in the pickup and SUV market, delivered exceptional driving experiences, and cultivated a group of loyal customers around their product offerings. To leverage the momentum and scale to more customers, Rivian announced mid-priced product lines, with R2 coming in 2026 starting from $45,000, and R3 in 2027.

As we see from the story of Rivian, positioning is foundational for businesses to stand out in a competitive market. With a differentiated product offering and a clear market position, Rivian effectively connected with a specific customer segment—environmentally conscious adventure seekers—and gained remarkable brand success.

Why Positioning Matters

What exactly is brand positioning? Positioning defines how your brand and product fit in the marketplace and in the minds of your customers.

"Positioning is the single largest influence on the buying decision. It serves as a kind of buyers' shorthand, shaping not only their final choice but even the way they evaluate alternatives leading up to that choice," Geoffrey A. Moore, author of *Crossing the Chasm*, explains.

David A. Aaker, the father of modern branding, says, "A brand position is the part of the brand identity and value proposition that is to be actively communicated to the target audience and that demonstrates an advantage over competing brands."[5]

How can you establish a positioning that solves unique customer needs and aligns with your product's strengths? It's about unpacking four important components.

- **Target Audience:** The customers the brand and product are for.
- **Category Convention:** The category in which the brand and product compete.
- **Brand Promise:** The unique value the brand offers to customers.
- **Reasons to Believe:** The primary reasons that support the promise.

A clear positioning is about finding your niche to establish a strong product-market fit. It is about connecting the customers you want to serve and the problems you solve for these customers with the unique products and services you create.

Today, we live in a highly commercial and overcommunicated society. In almost every category, customers have an abundance of choices. AI is accelerating product development, enabling rapid prototyping, and reducing time to market. Without a clear positioning, companies risk getting lost in the sameness, competing on features or price alone, and losing the opportunity to build unique value propositions and lasting connections with customers.

Take the car industry, for example. Simply being electric is no longer enough to stand out. Tesla had a first-mover advantage in the United States. It was once the only choice in the electric category, largely due to its innovative product and strong brand as a sustainability pioneer. Today, consumers have more options to choose from, from established companies such as Ford and Hyundai to new challenger brands like Rivian and Slate Auto, and across different price points. Imagine a future where all cars are electric. How do you differentiate? Rivian taps into a universal human desire—an adventurous spirit—and offers customers better ways to explore the world. In moving transportation systems entirely away from fossil fuels, the company also wants to change consumer mindsets and inspire others to change how they operate.

In China, the EV competition is growing rapidly as local brands arise, and Chinese consumers can choose from BYD, NIO, XPeng, Xiaomi, and others. In this increasingly crowded category, any new car brands entering the market need to have a clear differentiation in their offerings and create value and connections among the target audiences. Without a clear positioning, brands and products get lost in the market.

The startup world often uses the phrase "product-market fit" to assess whether a product can satisfy the needs of a market. Positioning is the foundational strategy to help new products and services find their places in the market and in the minds of customers. The strategic process helps clarify your differentiation, customer connection, and unique value proposition. To work toward your product-market fit, you don't need to wait until you go to market. You can start establishing and validating your market position during the building stage.

Positioning Is Your North Star

Many perceive positioning as a strategic tool for guiding marketing and communication only, but its true power lies in its ability to inform the brand and product offering and shape the business creation from the ground up. It's not just about how you talk about your product. It's about why you want to build something and what you want to build. It's also about choosing what you *won't* do.

Rivian's initial market entry shows how clear positioning can shape business and product strategy. By targeting adventure-seeking customers who valued sustainability and luxury, a market gap Tesla and traditional automakers left unaddressed, Rivian knew exactly what to build and, just as importantly, what to avoid. Strong positioning clarifies the category and business you are in.

For example, Airbnb isn't in the hotels business, but in the business of travel and experiences, offering unique stays ranging from tree houses to castles powered by a community of hosts. You can't be everything for everyone, so you have to choose your focus deliberately.

At its core, positioning starts with a clear vision, a defined market opportunity, and a customer problem to solve. This strategy serves as a North Star and guides decisions across the organization.

If you're an entrepreneur or you lead innovation, you may not realize that you need a positioning strategy. You may say, "Well, how can I form my product story?" Sometimes people think that the story is what they pitch to VCs or what they use in marketing and PR materials, but it can be more nuanced and emotional for customers. In searching for the right story, you need positioning to guide your messaging and make it consistent across different audiences. You should be able to return to the same story over and over and over, to align the product you build, the brand you create, and the entire organization.

Here's how your positioning strategy shapes what you build across three critical areas:

- **Product development:** Ensures products, features, prices, and design decisions align with customer needs and your North Star.
- **Brand development:** Guides naming, identity, packaging, content, and messaging to connect customers with your products and experiences.
- **Organizational alignment:** Reduces silos and misalignment through a shared vision.

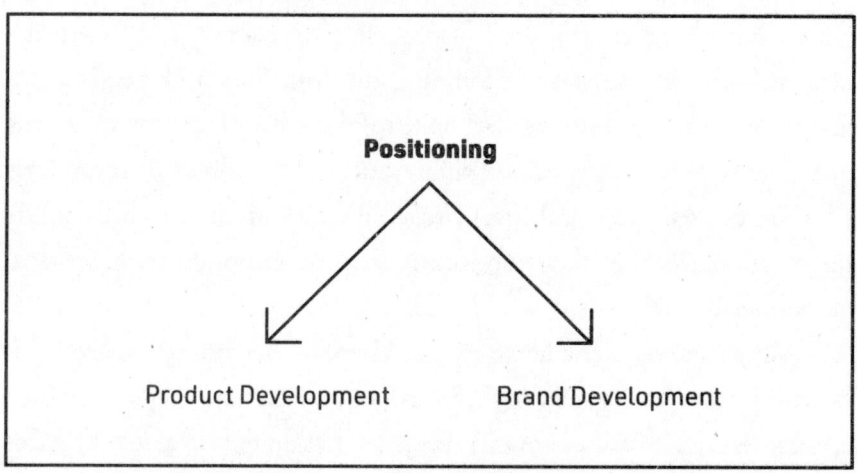

If your positioning and story aren't clear, internal issues will arise sooner or later. Aligning product and brand strategy is difficult, and most companies haven't figured out how to do this well. When teams only focus on delivering their own requirements without a shared strategy guiding the execution, that's where disconnections happen. This is a common struggle. I've seen this happen repeatedly where teams often skip the entire positioning strategy process and jump straight into execution.

Avoiding Common Positioning Pitfalls

We'll illustrate three common pitfalls of brand positioning and discuss how to avoid them. If you don't lay strong foundations early on, you are more likely to waste time on building the wrong products and struggling to attract customers, resulting in go-to-market failures and scaling challenges.

A Focus on Technology over Customer Needs

The most common mistake is focusing on what you do with your new products, features, and technology, instead of addressing real customer needs. For tech products, capabilities are important because they determine what customers can do. But obsessing over innovation and technical advancements instead of building customer benefits often leads to failure. As Harvard Business School professor Tom Eisenmann reveals in his book *Why Startups Fail*, which draws on hundreds of interviews with startup founders and investors, companies often rush into building solutions without fully connecting them to customer problems and unmet needs.

We see a similar challenge in the AI space. According to RAND, a nonprofit research organization, more than 80% of AI projects fail, which is twice the rate of failure for tech projects that do not involve AI.[6] After interviewing 65 data scientists and engineers, researchers identified that the top issues are the misunderstanding and miscommunication about the exact problem AI is trying to solve. They tend to focus more on the greatest technology rather than real customer problems.

This technology-first approach reflects the broader AI race driven by the transformative potential and the fear of falling behind competitors. As companies scramble to find viable consumer use cases, what Sequoia calls "AI's $600B Question"[7]—the gap between massive investment and actual revenue—highlights why consumer insights and positioning strategies are essential for AI products to succeed.

If you're developing an AI-enabled product, it's important to shift the mental model from "we can't fall behind" to focus on "why it matters to customers." The simple questions around who it's for, why it matters, and how it's different can lead you to find your unique position in an increasingly crowded field.

I've learned this important lesson from Steve Jobs: "You've got to start with the customer experience and work backwards to the technology. You can't start with the technology and figure out where you're going to sell it."[8]

This philosophy shows that Apple's approach to product development starts with customer benefits. It's not about leading with technology or features. This vision has guided how Apple develops a cohesive ecosystem of hardware, software, and services that provides customer value. This may seem obvious, but it's difficult to execute in daily practice. That's why it's essential to constantly ask yourself these fundamental questions. Many of these failures are preventable, and many of these lost values can be gained if you put enough effort into the foundation. Every founder knows the sleepless nights spent wondering, *What if nobody wants this?* This is what I mean by winning from day one: It's really about *not losing* from day one and preventing the costly mistakes that come from poor positioning.

Mass Appeal Without Purposeful Focus

Another common challenge I've seen is that companies try to serve everyone for their new products, resulting in a generic targeting and a lack of differentiated value proposition to connect with core audiences. This may not lead to massive failures like the first pitfall, but it makes everything you do less effective.

Even a successful brand like Apple didn't always get it right. When the first Apple Watch launched in 2016, Apple positioned it as "the most personal device" at the product launch event. However, at a time when the iPhone was already ubiquitous and personal, the message didn't resonate strongly. People bought Apple Watch for health and wellness reasons. Over time, Apple evolved the positioning from "the most personal device" to "the ultimate device for a healthy life," focusing on customers' motivations to stay active and healthy.

This strategic focus opened opportunities to create more compelling marketing campaigns around personal health and wellness and communicating hero benefits and use cases to health-conscious customers. This shift not only aligned benefits with customer desires but also fueled new services like Fitness+ to further deepen customer engagement.

If you're launching something new, there's a temptation to go after a big market, aiming for everyone, but every successful product begins with a niche customer segment and evolves to reach the mass market over time. Apple started with Macintosh for creative minds. Amazon started with selling books online before it became an everything store. Having a purposeful focus is more effective in the abundant and fragmented world we live in today.

Fail to Stand Out in the Sameness

Another positioning challenge is a lack of product and brand differentiation. The startup world often uses the metaphor of "vitamin vs. painkiller" to assess the potential of new products. A vitamin is nice to have, providing incremental improvement, while a painkiller addresses a specific pressing problem and delivers significant value. Successful products are painkillers—solving real customer pain points rather than offering incremental improvements.

Today, streaming services face a differentiation challenge. The other day, I chatted with my friend Jay about all the streaming services our family has. We have Hulu, Disney+, Netflix, Prime Video, and Apple TV+. The experiences among these services aren't too different. If you remove their branding, the main difference becomes the content. While branding is critical for differentiation and emotional connections, it cannot fix fundamental issues in product value, functionality, or quality. When online streaming first entered the market, Netflix was innovative and disrupted the traditional content consumption model. But now, streaming technology is simply table stakes. How can you continue to differentiate? It requires continuous innovation on content offerings across different genres for their fans and building their unique value propositions, like how ESPN built its reputation around sports.

Positioning is not a one-and-done exercise because the business, the category, and customer needs will all change. The strategy that worked in the early growth phase may no longer work for the scaling phase. As you can see, even successful brands need positioning and repositioning because of the dynamic market environment and changing customer behaviors and expectations.

To avoid these common pitfalls, you must go beyond product features to understand what truly motivates your customers and build better and more differentiated products.

Turning Insights into Strategic Opportunities

Effective positioning requires a deeper understanding of customers' problems and designing solutions around them. To do this well requires consumer insights and human creativity. It's about identifying an insight and turning it into a strategic opportunity that fuels product innovation and brand development. Understanding the cultural and social context is also important as the cultural lens gives you fuel and helps you stay relevant.

Uncovering True Customer Insights

Being customer-centric sounds obvious. Of course, every company and everyone wants to be customer-centric, but true customer obsession is hard to achieve. Solving customer problems is like peeling an onion. Each layer you uncover brings you closer to the truth—the root issue. However, many products only address surface-level issues, missing the deeper, underlying customer needs.

Different from facts and information, an insight can be a catalyst for transformative ideas and strategic opportunities. An insight may come from your understanding of customers, from your category analysis, or from the vision of your product. An insight is a revelation that helps you see things differently and frames your problem in a whole new way.

How can you uncover deeper consumer insights? Think like an anthropologist. Walk in customers' shoes with empathy. Data is not insights. When data becomes so pervasive and accessible, we have to make sense of the data with human empathy and intuition to gain deeper insights beyond the surface.

Consider the consumer insights that drive the design of Airbnb. In the early days, Airbnb started as a low-cost alternative housing service, primarily targeting conference attendees. But this changed in 2014 when Airbnb did a repositioning and rebranding to evolve its brand and product beyond the initial budget accommodations and deepen customer connections.

Airbnb tapped into an insight that a new generation of travelers had a strong emotional desire and aspirational needs for unique experiences. Seventy-seven percent of Airbnb guests said they chose Airbnb because they "want to live like locals." Thirty-six percent of Airbnb bookings are for guests ages 25 to 34, and 23% are for guests ages 35 to 44.[9] The needs for millennials and Gen Z travelers are different from baby boomers and Gen X. Airbnb evolved itself to become a community-driven brand for magical travel experiences, emphasizing unique travel for adventurous travelers, families, business travelers, and anyone who wanted to explore the world, not just find a place to stay.

As part of the new positioning, Airbnb redesigned its brand logo, storefront, technology, and the entire digital experience. This strategic shift elevated the Airbnb offerings beyond a low-cost price play to focus on unique stays. It also opened new opportunities for the company to tap into a new generation of travelers who were willing to pay a premium price for unique experiences. As reflected in its product offerings, Airbnb partnered with hosts to curate and offer experiences from tree houses to castles.

Looking back, by uncovering the deeper emotional desire for authentic experiences rather than affordable accommodations, Airbnb transformed its business and created its own category. Airbnb has both emotional and functional benefits in its positioning that resonate with a new generation of travelers.

Take social media as another example. Platforms like Twitter (X), Facebook, and Instagram dominated social media for years, until Bluesky identified an underserved segment: users who want algorithm-free experiences and greater control over their data and feeds. The underlying customer insight was that many users felt frustrated by algorithmic manipulation and wanted more authentic connections and community-based conversations.

Launched as an invite-only beta service in 2023 before opening to the public in 2024, Bluesky positioned itself around a decentralized, open, and user-centric social platform. In January 2025, Bluesky reached 30 million users in just over one year after its public launch. As one user perfectly captured the positioning, "This is my social media platform now . . . it's NOT constant ads, 'viral' clickbait bs, no moderation tools or controls & an algorithm built to incite anger."[10] By giving users control over their experience, Bluesky demonstrates how clear positioning around frustrated customer segments can create new market opportunities.

Tools to Uncover Insights

If you ask different teams in your organization who your customers are, you'll be surprised to find out that you often get different answers. This lack of shared understanding of your customers is a concerning sign: It indicates that your internal efforts might be disjointed and fragmented. Helping the entire organization better understand your customers is essential for developing a shared positioning strategy and building a successful business.

Three tools are particularly useful to gain insights and intelligence and validate and inform your thinking: customer segmentation, customer personas and profiles, and competitive benchmarking. Each tool can help you uncover deeper insights to strengthen your positioning. These tools can help identify your ideal customer segments, uncover their unmet needs, and understand how they are served by competitive offerings.

Customer Segmentation

I've learned from my past work that customer segmentation is one of the most foundational studies to inform product development, positioning strategy, brand development, and go-to-market planning. Customer segmentation provides the size of the total addressable market and customer segments with distinctive buyer personas. The study is a truth-teller and helps you define your ideal customers (including those who are not your customers yet but could be in the future). This process isn't just a research exercise; it ensures that your positioning is rooted in real customer insights, enabling you to create products and experiences that resonate with your audience.

Rivian started by targeting the pickup and SUV market and catering to the environmentally conscious outdoor adventurers. By focusing on this customer segment, Rivian was able to define a clear target audience and tailor its offerings specifically to their aspirations. This strategic focus not only allowed Rivian to stand out in a crowded market but also build a strong brand around adventure.

As Rivian continues to scale, the company looks to expand and reach more people with more affordable models. One product rarely serves everybody. Different models, features, and price points are required to appeal to unique customer segments. Customer segmentation is a tool to help you clarify your focus. A more refined target segment gives you a clear direction to build your brand and product experience.

Customer Personas and Profiles

Customer personas and profiles are useful for visualizing your target customers and guiding product and brand design. Customer segmentation offers you a high-level overview of your ideal customer segments, while personas give you qualitative and vivid insights into your customers.

The qualitative studies are complementary and provide deeper understandings of each customer type. This work is particularly useful for designing customer experiences, developing features, and creating marketing campaigns because you already have a clear customer in mind.

When you deeply understand your customers—like a working mom, Anna, from Michigan or a retired adventurer, Bob, from California—they are no longer just numbers and customer segments. You humanize the customer you aim to serve. You can then build a strategy around their needs and desires.

Competitive Benchmarking

Gathering intelligence about your competitors and the market environment is not competitor-obsessed, but rather understanding how well your potential customers are served today and identifying unmet needs.

Competitive benchmarking provides a clear picture of where new products and innovations sit in the market. How can you differentiate your product offering from the alternatives? The category understanding helps define the market opportunity, identify your differentiated niche, and better position your product in the marketplace and in customers' minds. In other words, competitive analysis is a strategic input to develop effective brand positioning.

But when you rely heavily on benchmarking, it can easily become a comparison and a checklist. You must think outside the box and challenge the conventional way of benchmarking to find opportunities. Today, with A/B testing tools widely accessible, it's easier to gather data and customer feedback, and you might react to this feedback with quick assumptions: "Customers don't like this, so let's change it." But the clicks and conversions may not tell the full story. It's always important to ask, "Why don't they like it?" Digging deeper and getting to the "why" requires more curiosity and thoughtful questions, which can lead to true insights and your "aha" moments.

Many startup ideas may begin with a hunch, an inspiration, or an instinct. To avoid being product and feature obsessed, Amazon uses a PR/FAQ and a working backwards mechanism. For any new initiatives, Amazon asks teams to start with a hypothetical press release that frames a new product idea from the customer's perspective. This process empowers teams to think deeply about customer problems, understand how their new product idea solves them, and work back from all necessary elements to support an actual launch as if it's already launched to the public. As Jeff Bezos put it, "The working backwards process is a huge amount of work. But it will save you even more work later."[11]

If you're innovating and seeking your next breakthrough, these customer-focused tools can help you identify your meaningful opportunities that transform how you develop products and how customers experience them.

Why Human Insight Still Matters in an AI World

Positioning is about finding your unique strength in the market and in customers' minds, which requires consumer insights. With AI tools promising instant answers, many people question the value of traditional research studies and consumer surveys. Companies are increasingly leveraging AI to run research and eliminate insights-related jobs as an easy cost-cutting measure. While AI can serve as a great supplemental tool, it cannot replicate (at least for now) the human ability to interpret data and translate that data into actionable insights.

Quick Information vs. Strategic Insights

AI accelerates information gathering, offering suggestions that make the insights process more efficient and accessible to us. However, there's a big difference between information and insights. It's easy to confuse convenience and efficiency (which AI can offer) with effectiveness.

The key question to ask: Is the information we're getting via AI substantial and valuable enough to drive strategic decisions? You need a strategy leader who can sift through information, interpret results, understand human perceptions and behavior, and consider business context to develop effective strategies. There's no shortcut from data to actionable insights. It requires human interpretation and decision-making.

Another disadvantage of heavily relying on AI tools is that they recycle existing knowledge and internet data to produce results. Since everyone is now accessing the same handful of AI tools and data sources, there's a real risk of homogenous thinking. When two similar startups prompt AI for positioning guidance, they'll likely receive similar outputs—the new conventional wisdom.

You need the strategic ability to determine what's valuable for your specific product or your company. A positioning approach that works for Company A won't necessarily benefit Company B. It takes human analysis and judgment to uncover fresh, unexpected insights that unlock new opportunities. Otherwise, you can fall into the trap of conventional thinking without meaningful differentiation. Without skilled professionals to filter data and construct proper surveys or prompts, you risk generating misleading outputs based on the wrong data.

This challenge reminds me of the history of performance marketing, which we discussed in earlier chapters. You can easily track views, clicks, and conversions from Google Search or Facebook Ads, but their long-term effectiveness is much harder to evaluate. When every company adopts the same playbook, they are likely to focus on the same metrics and draw the same conclusions as everyone else.

While AI offers tremendous promise, it's important to remind yourself that your goal isn't just process efficiency—it's creating happy customers. Imagine a new world where AI generates your positioning, insights, logos, websites, copy, photography, videos, and code. If everything collapses into generic and conventional sameness, how will you stand out? How will you *really* win the hearts and minds of customers?

Operationalizing Strategy Across Your Organization

For the positioning strategy to have the biggest impact, you have to use it beyond the marketing team and align with the entire organization. As we discussed earlier, brand positioning serves as both a product and business strategy, which is why you can't undertake the positioning exercise alone. Inviting cross-functional partners into the process is essential to create a shared vision.

Input from product and business teams ensures the strategy aligns with the product and broader business vision. Sometimes, your product may not be able to deliver the positioning immediately, but it sets a North Star for the product to iterate and improve. The collaborative process will invite meaningful conversations and debates. In this process, marketing has a seat at the table. They bring the voice of customers and deeper insights into meetings and engage in difficult conversations and debates.

When your new product is still in a fluid stage, timely discussions are essential to ensure the positioning evolves alongside development. Sometimes companies outsource the positioning strategy, but the journey, conversations, debates, and nuanced organizational challenges are not easily outsourced. Without a collaborative process earlier, you might struggle to "sell" the strategy internally and implement it effectively later.

Once you have a shared vision and positioning strategy, it's vital to ensure everyone on the team understands it, as well as using it to guide the execution and deliver better customer outcomes. To think big, you must develop mechanisms to operationalize your strategy across the organization. An operational playbook is as important as developing a strategy. Having an aligned approach helps you set a consistent and unified vision with product and customer experience, reducing the risk of moving in different directions and saving time and resources on unified initiatives.

Additionally, having an aligned strategy also helps avoid the risk of needing a repositioning effort later if your product launch fails to gain traction or the business becomes stagnant. You can always do a repositioning and rebranding, but consider the entire process to get it right in the first place, as it's often more costly and time-consuming to manage a repositioning.

Most companies use their mission statement to guide their execution, which we'll explore in chapter 4. It's a foundation for the long term, but sometimes it can feel too high-level to guide day-to-day goals. The positioning strategy directly guides your execution and is a part of the tangible thinking behind the mission and the story. To galvanize the entire company, you have to operationalize your positioning strategy and ensure it applies to different business functions across your organization.

40 Questions to Ask About Your Positioning

To operationalize and implement strategy, here's a list of 40 questions you can ask about your positioning and how it applies to different business functions across your organization.

For Business Strategy

Positioning clarifies the category and business you are in. It crystallizes your business focus and guides where the business is going and where to invest.

> Does our organization have a positioning strategy?
>
> What is our positioning strategy?
>
> Which business are we in?
>
> Who are our target customers?
>
> How do we offer distinct value to customers?
>
> Do we want to serve a niche market or a broader customer base?
>
> How can we integrate positioning strategy effectively across our organization?
>
> Do we need a repositioning? Why?

For Product Development

Positioning provides a North Star for product development and acts as a strategic filter to help you make strategic decisions, ensuring the product roadmap meets customer needs and aligns with the brand.

> What is our product vision and story?
>
> What customer problems do we try to solve?

What insights do we have to support this decision?

How does this new product or feature add value to customers?

Do we need this new product or feature?

Which product features should we prioritize?

Is our product essential to customers and aligned with the core brand promise we deliver?

What's more essential to our core customers?

For Brand Development

Positioning guides how you deliver your brand naming and identity, design experience, and product packaging and content. It also informs your tone and voice, your brand's personality, and how it shows up to consumers.

What's our brand strategy?

Where do we fit in this world?

What does our brand stand for?

What is our brand vision and story?

Why do we matter?

Do our name and identity represent who we are?

Is this creative execution ownable to our products and services?

Why would we need a rebranding?

For Go-to-Market

Positioning streamlines the process of bringing a new product to market and informs and guides your go-to-market and sales strategy, ensuring all activities align with an overarching strategy.

How does our positioning strategy support our GTM activities?

What is our unique positioning, value proposition, and messaging for go-to-market?

Which customer insights are most critical in shaping when and how we launch?

What are the channels to reach our target customers?

How do we achieve product-market fit?

How do we sequence our GTM efforts to reach our target customers?

What partnerships could strengthen our market entry and support our positioning?

Are our sales teams equipped with the tools to articulate our positioning clearly?

For Customer Service

Positioning helps the customer service team understand who they serve and how best to answer their questions.

Which of our customers most require customer support?

Are we addressing different customer pain points effectively?

How do our customer service protocols reflect our brand, promise, and tone and voice?

How do we collect feedback from customer service to refine our positioning?

Are we consistently reinforcing our promise through customer service interactions?

How should onboarding strategies differ for various customer segments?

What tools and resources do our customer service teams need to communicate our brand, products, and services?

How can we turn customer service interactions into relationship-building moments?

——————————— **Chapter Summary** ———————————

Positioning isn't about what you sell. It's about establishing a unique value proposition to help customers buy. Positioning is your differentiation and the truth of your product in the mind of customers.

A clearly defined positioning includes four important elements:

- **Target Audience:** Who the brand and product are for
- **Category Convention:** The category in which the brand and product compete
- **Brand Promise:** The unique value the brand offers
- **Reasons to Believe:** The primary reasons that support the brand promise

Positioning is your business and product North Star. At its core, positioning starts with a clear vision, a defined market opportunity, and a customer problem to solve. This strategy serves as a North Star and guides decisions across the organization.

Avoid three common positioning pitfalls. These pitfalls include focusing on technology over customer needs, trying to serve everyone without purposeful focus, and failing to differentiate and stand out in the sameness.

Effective positioning requires consumer insights and human creativity. True customer obsession requires going beyond the surface to get closer to the truth, the root issue, and tapping into deep emotional desires and aspirational needs.

Identifying an underserved customer segment and underserved needs allows you to build a new product and a new brand around them. Customer segmentation, customer personas and profiles, and competitive benchmarking are three essential tools to gain insights and intelligence.

Human insight still matters in an AI world. It takes human analysis and judgment to uncover fresh, unexpected insights that unlock new opportunities. Without this human element, you can fall into the trap of conventional thinking without meaningful differentiation.

Operationalizing strategy across your organization. For the positioning strategy to have the biggest impact, it must go beyond the marketing team to align with the entire organization.

Ask 40 questions about your positioning. These questions apply to different business functions across your organization such as business strategy, product development, brand development, go-to-market, and customer service.

Chapter 4

Establish Your Brand Identity

From Creative Execution to Value Creation

After you've clarified your new product's positioning, how can you bring the strategic vision to life? Branding gives your product meaning and differentiation, creating intrinsic value that will resonate with customers.

From crafting your brand story to developing naming and visual identity, this chapter explores the essential foundations of establishing your brand identity.

How Irreverent Branding Made Liquid Death a Billion-Dollar Company

Bottled water is one of the most commoditized products. So how did canned water branded with a melting skull logo become a company valued at $1.4 billion?[1]

In just five years after its 2019 official launch, Liquid Death turned one of the most mundane products into a cultural phenomenon with avid fans. Its viral popularity surprised the entire industry, proving the incredible power of brand identity when deployed with creativity and boldness.

Most bottled water brands position themselves in contrast to tap water, largely focusing on being "pure" or "clean" as key differentiators. The unique selling points of premium water brands such as Evian, Smartwater, and Fiji center on water source location, pH balance, or mineral content. For example, Evian emphasizes *"natural spring water directly from the French Alps."*[2] Fiji focuses on "earth's finest water" from its ancient artesian aquifer and a pH level of 7.7.[3]

Mike Cessario, CEO and cofounder of Liquid Death, had a simple idea—making the world's plainest drink less boring with an irreverent, humorous approach to branding, no-plastic packaging, and marketing. The idea of Liquid Death was born, first without a product. To test the concept, Cessario first launched a video to gauge interest on Facebook, which received three million views[4] before the water was even available for anyone to purchase. This social traction helped Liquid Death raise investment capital and build a startup to create and distribute the product.

With the provocative tagline "Murder Your Thirst" and the hashtag #DeathToPlastic, Liquid Death began selling its products on its website in January 2019 as a direct-to-consumer business. Every detail from brand identity, packaging, marketing, and content to the customer experience was cohesively irreverent and rebellious. Even the language around the sustainability of their recyclable aluminum cans was wrapped in humor.

Initially, Cessario thought Liquid Death would be a niche product only for concertgoers seeking a healthier alternative to alcohol or energy drinks, but he greatly underestimated his potential market. It turned out that there's a big market for this. Many young people loved the edgy feeling, humor, and emotional connection they discovered in Liquid Death.

Organic word of mouth accelerated the brand's rapid growth. Because of its unexpected and memorable brand identity, early adopters happily endorsed the product on social media without any paid promotions.

Liquid Death rocketed from $2.8 million in sales in 2019 to $333 million[5] in 2024. The company expanded beyond water into soda-flavored sparkling water and iced tea and added distribution partnerships with major retailers like Whole Foods and Target, in addition to its original direct-to-consumer e-commerce. Since day one, Liquid Death had brand power built in; it's as much a lifestyle statement as a water company. As Cessario said, "We want to actually entertain people and make them laugh in service of a brand. And if you can do that, they're going to love your brand because you're giving them something of value."[6]

In the story of Liquid Death, the product is a barely differentiated commodity, but the brand is totally unique. It's proof that a distinctive and powerful brand can attract customers organically from day one, creating a successful company from the ground up.

What Tech Leaders Can Learn from Liquid Death

Unlike water, tech products and services are usually not commodities. They have distinctive product features and engineering specifications that need to be explained and promoted to potential customers. The big problem arises when those priorities block new tech brands from establishing emotional and cultural connections with potential customers. The brand can become part of your innovation. Liquid Death's success is a direct reflection of customer behaviors and how customers make choices—not through rational reasoning but emotional and cultural connections.

Effective branding is rooted in a deep understanding of human psychology. From typography to color to messaging, each design choice is intended to influence customer perceptions and behaviors. This is why fashion and luxury categories thrive. People know they're paying more for a Louis Vuitton handbag or a Balenciaga hoodie than they would for an almost-as-good generic product, yet they willingly pay that extra premium. The brands we love become part of our identity, reflecting who we are, our personal values, and the images we want to project to the world.

If you show or tell a new acquaintance "I drive a Rivian" or "I shop at Whole Foods" or "I drink Liquid Death," you're not just describing your consumption habits—you're declaring who you are.

Brand identity isn't just creative execution. It's a strategic lever for differentiation and business value creation. Branding gives your product meaning and differentiation. It is an intrinsic part of the product experience to connect with customers. Without it, your product becomes generic and doesn't stick in customers' minds.

In his book *Design-Driven Innovation*,[7] author Roberto Verganti explains product meaning and language:

> Products appeal to people and their needs along two dimensions. The first one is familiar to anyone managing innovation: The utilitarian function, provided by product performance and based on technological development. The second dimension concerns sense and meaning. It's the "why" of the product—the profound psychological and cultural reasons people use the product. This dimension can imply an individual or a social motivation.

Liquid Death focused overwhelmingly on the second dimension, while most tech products focus overwhelmingly on the first dimension. Long-term value depends on finding the right balance for your product.

Remember Ring's founding story from chapter 1? Its rebranding from Doorbot to Ring provides critical lessons on how early branding efforts created value for new ventures. In 2014, when Ring had no products beyond the Doorbot, Simon Cassels, who became the CMO, explained how he used a "storyboard" to make the vision and product roadmap tangible:

I created modest yet sufficient visualizations of the "Rings of Security" surrounding individuals, their homes, and neighborhoods, thereby imparting tangibility. I used an image of a soap dispenser to represent the first Ring Video Doorbell. Illustrations depicted an ecosystem of cameras in and around homes, alongside the concept of a social network for neighborhood security, which would later materialize as the Neighbors App.[8]

This conceptual foundation helped Ring launch, evolve, and scale to become one of the largest home security brands.

Both Liquid Death and Ring demonstrate that branding isn't just about marketing. It's about creating differentiation and competitive advantage that can guide product development and create business value from day one.

Branding Foundations

Three crucial strategic decisions shape an overall brand foundation: brand positioning, vision and story, and name and visual identity. These are strategic decisions that shape overall brand foundation.

First, you must position your product and brand in the market and in the minds of potential customers. We've already covered this topic in chapter 3. Your positioning strategy can drive and guide both brand and product development. Identify the North Star that defines why you exist in the world and bake it into your product and organizational DNA from the start.

Second, your vision and story must bring your positioning to life in a narrative format that feels clear and tangible. Your story conveys the meaning of your company and your products. It can be told across different channels and environments. It's simultaneously your message to customers on why they should choose your product over others, your pitch to investors on why they should invest, and your narrative to employees on why they should join the journey.

Third, your name and visual identity must become key assets that embody the brand and appeal to customers. Customers can see it, feel it, and use it. Your name and visual identity show up inside and outside the product. They show up on online and offline channels to connect with customers and leave an impression. They are the most important branding decisions for your company and products. They can't change often, so make sure they can grow over time with the company as valuable assets.

These three foundational elements, coupled with visual system, tone, voice, and personality, work together to express the brand, evoke feelings, and form customer connections.

Bringing Your Vision to Life Through Story

Every great brand and product has a story. In fact, customers don't just buy products; they buy stories. The human mind connects to stories better because we process information through stories, frames, and patterns.[9]

According to George Lakoff, a cognitive linguist and philosopher, "Human thought processes are largely metaphorical. This is what we mean when we say that the human conceptual system is metaphorically structured and defined. Metaphors as linguistic expressions are possible precisely because there are metaphors in a person's conceptual system."[10] It's built into how we think and make sense of the world.

This insight explains why you need stories to convey your vision, to make your product roadmap tangible, to effectively connect with customers, and to help people see the connection and potential.

Brand and Product Integration

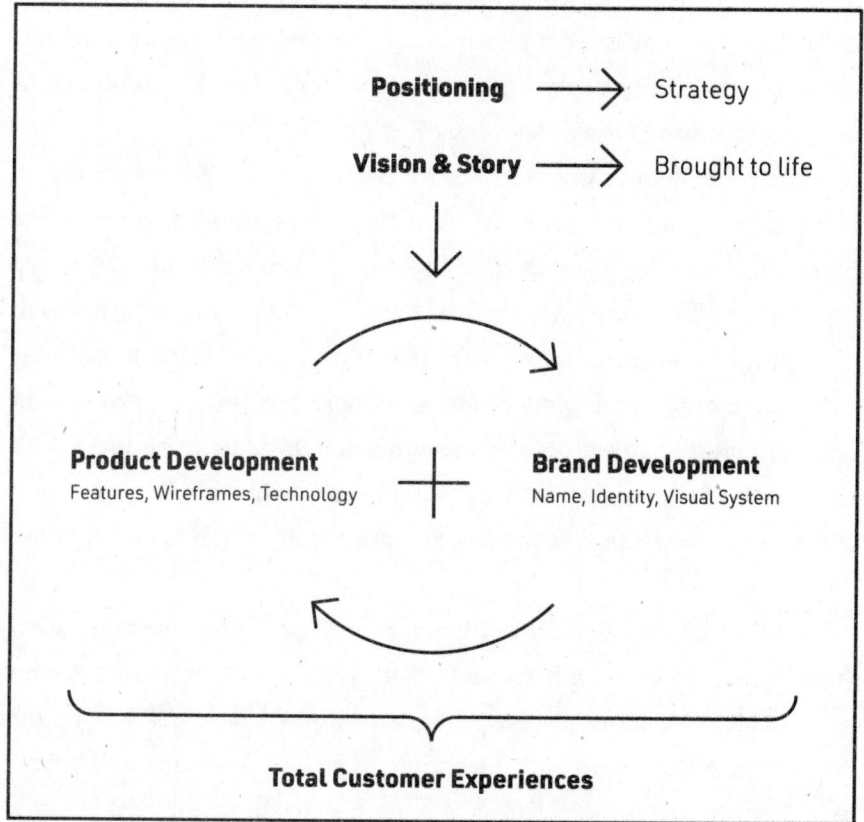

Total Customer Experiences

Start with Why

Your brand story articulates your point of view on the company's existence in the world and lays a strategic foundation for why you do what you do. You start with the fundamental questions:

- Who are you?
- Why do you matter?
- Why are you important to people?
- Where do you fit in this world?

In the early stages of building a new business, there are countless demands on time and resources. When the mindset is focused on survival, your story could seem like a luxury that can safely be deprioritized until later. In reality, clarifying your brand story plays a critical role in new product development and design.

Great companies like Airbnb, Rivian, and Spotify embed passion and purpose into their products. Airbnb goes beyond being an online home rental marketplace to foster a sense of belonging. Rivian is not just another EV company; it refines adventure. Spotify isn't just a tech platform, but an entertainment brand with a passion for music culture. These emotional connections transcend technology, showing up across product, marketing, and customer touchpoints to connect with people's hearts and minds. Your story—your why—drives the brand and product development process and guides all product and branding decisions.

Xiaomi, a Chinese electronics brand, exemplifies how having a clear "why" from day one can help you break through in a crowded category and enable long-term success. Founded by Jun Lei in 2010 with the mission to make high-quality technology widely accessible and enjoyable, Xiaomi had an idea to make smartphones more affordable while maintaining high-quality experiences. It positioned its brand in contrast to the premium smartphone brands such as Apple and Samsung, promising similar features at far lower prices. The brand name Xiaomi means "small rice" in Chinese. It was chosen to reinforce the company's humble values and broad accessibility. To prove this commitment, Xiaomi made an unprecedented promise: The company would maintain operational efficiency to reduce costs for its users and never earn more than 5%[11] profit from hardware sales. This community-first approach helped build a passionate global fan base, propelling Xiaomi to become the world's third largest[12] smartphone brand while also expanding into a broad range of innovative tech products and services.

Balance Bold Vision with Reality

While vision and emotional connection give your product meaning, they must be grounded in reality to build credibility. A vision and narrative without a product to deliver can lose trust.

Let's look at a few examples that exemplify the power of a clear vision and narrative and clearly articulate why they do what they do.

Stripe: To grow the GDP of the internet.
Figma: To make design accessible to everyone.
Robinhood: To democratize finance for all.

These visions succeed because they're both bold and actionable. Stripe's payments infrastructure enables companies to transact online, serving as a backbone for global commerce to grow the GDP of the internet. Figma's web-based design tools empower non-designers to collaborate, fundamentally making design accessible to everyone. Robinhood disrupted traditional brokerages with commission-free trading, removing barriers and democratizing finance for everyday investors.

Conversely, there are also many examples of empty promises that weren't grounded in real customer needs or desires, which ultimately contributed to the failures of these companies.

WeWork: To elevate the world's consciousness.
Humane AI: To build a future where AI seamlessly integrates into every aspect of our lives and enhances our daily experiences.
Quibi: To create a new category of short-term entertainment for mobile devices.[13]

WeWork's promise was too broad and didn't connect closely with what a coworking space solution actually delivers. Humane AI communicated a philosophy for the future of AI but didn't anchor it in real customer benefits. Quibi promised to revolutionize mobile entertainment with premium, short-form content, but completely misread how people consume video. Despite raising nearly $1.75 billion, it shut down in less than a year.

The lesson here is that a strong brand story must work in harmony with the product you are developing and balance aspirational vision with market realities, grounded in authentic customer needs and product truths rather than technological possibilities.

Peter Thiel's book *Zero to One* makes a similar point: "A company has a monopoly on its own brand by definition, so creating a strong brand is a powerful way to claim a monopoly." However, he also cautions, "Beginning with brand rather than substance is dangerous . . . No technology company can be built on branding alone."[14]

Similarly, David Ogilvy, known as the Father of Advertising, advocates the necessity of building a great product: "Great marketing only makes a bad product fail faster. It is only when the product is good that marketing can help build a strong brand."[15]

Keep in mind that your long-term vision may stay the same, but how you tell your narrative may evolve over time as the product and business move from one stage to another. What worked at the early stage may not be relevant to the scaling phase. You must regularly assess and update your narrative to be relevant to customers and their experiences while staying true to your product and business.

Naming Is a One-Way Door Decision

Every new product and company needs a name, and those are the single most important branding decisions you will make. Names are not only critical; they are usually irreversible decisions for any product or company.

Most people don't realize the importance of naming until it's too late. Once you launch a name publicly, it's hard and costly to change. From the product experience to marketing materials to PR announcements, your name becomes the anchor for your entire customer journey. For example, even though Facebook changed its parent company name to Meta and Twitter changed to X, most people still call them Facebook and Twitter. Even big and successful companies struggle to get customers to use their new names.

A great name aligns with your product's value, represents where the business is going, and appeals to your core customers. A great name is also clear, easy to understand, and flexible enough to make sense in the future as your business grows. It reduces customer acquisition costs and simplifies customer decision-making by being memorable and distinctive. Think of it as a mental shortcut for customers.

However, naming is one of the most difficult workstreams as it can be subjective and nuanced. Because it's a one-way door decision, you must invest enough time, rigor, and attention to do it as well as you possibly can. The danger of not getting a name right at the outset means that failure is not an option.

When I led the 2023 launch of RxPass, a Prime benefit from Amazon Pharmacy, we went through a rigorous naming process. This subscription service charged $5 per month to give Prime members access to more than 50 of the most common generic medications. I remember vividly the two biggest questions we debated:

Should we include the word "Prime"?

Should we include the word "Pharmacy"?

At that point, Amazon Prime was a well-known brand with very strong brand equity. Prime members loved the service. But, if we branded RxPass under Prime, it would lose its immediate connection with Amazon Pharmacy, a new brand and category we were working hard to establish.

We explored many options, stress testing how we might launch and communicate those options and how customers would respond. We developed mock-up materials for every name option, covering the end-to-end customer journey from discovery to sign-up to engagement. This process helped everyone see the pros and cons of the major name contenders, making it easier to drive alignment toward the final choice. We landed on "RxPass from Amazon Pharmacy"—or just RxPass for short—aligning with the actual customer experience and making it distinct from Prime.

Good names give immediate meaning and understanding without explanation. While finding the right name that's legally available and creatively appealing is challenging, a poor name or a made-up name requires significantly more marketing investment to build understanding—investment that could be allocated to customer acquisition instead. This represents a type of invisible waste that many startups don't recognize until it's too late to change course.

The Amazon Pharmacy example reinforces the lesson that naming is both a strategic and a creative exercise. You can use your positioning strategy to guide exploration of potential names and choose one that captures your brand's essence and resonates with customers. Since this is a one-way door decision, you should gather customer feedback through research and user testing and get input from your trademark team before finalizing it.

Brand Identity Is a Strategic Long-Term Asset

Your brand identity is a long-term investment that grows in value over time. When the designer David Turner met with Jeff Bezos in 1999 and delivered Amazon's smile logo, he designed it to reflect two business needs at the time: customer-centric service and an "everything store" that was branching out far beyond books. The arrow connected the A to the Z within the word "Amazon." It was a simple and versatile design to grow with the company over time—enabling it to literally sell anything from A to Z. As Amazon expanded its products and services into new categories, the logo evolved slightly, but its core elements stayed the same to preserve Amazon's brand equity.

As Turner explains:

You shouldn't think of identity design as a marketing expense. Like other assets a business invests in, well-designed visual assets deliver value long after they are paid for, benefiting a brand for decades with no additional cost. Think of the Amazon logo we designed. Billions of boxes delivered over two decades, and every one of them with a smile on its face. That's an idea that will never grow old. And it works on everything from a packing slip to a fleet of aircraft.[16]

Brand identity is more than a name or logo. It includes tangible and intangible elements, an entire visual system, tone, and voice to connect with customers and communicate your value proposition. These foundational elements are used in product design, UX flow, website, packaging, marketing, customer services, PR outreach, and *all* customer touchpoints to influence perceptions and behaviors. A strong, distinctive brand identity aligns with your vision, story, and product experience. Then, as you go to market, it is communicated and reinforced through company activities.

As David A. Aaker explains, "A brand identity should help establish a relationship between the brand and the customer by generating a value proposition involving functional, emotional, or self-expressive benefits."[17]

This is why a brand identity is not just a creative exercise but a strategic, long-term asset. A unified identity across product and brand experience is essential for creating lasting customer connections. This is why brand designers and product designers should come together to create cohesive customer experiences in tandem. (In chapter 5, we'll see how this works.)

Branding projects involve lengthy legal and trademark processes. To launch new products and services, you need many options to get through legal and trademark clearance. This is another reason why you must balance agility and speed with careful thought and planning.

Brand Architecture: Company, Products, and Features

To develop an effective brand identity, it's essential to understand the hierarchy of your brand, products, and features and how they relate to one another. When you name or brand a new product or feature, you might not think too far ahead because "it's still in an early stage." But every naming and branding decision influences both your current and future offerings, as well as how customers will perceive your brand.

A consistent brand architecture can radically simplify your messaging to customers, make your marketing investment more efficient, and reduce potential fragmentation and complexity. Unlike mature companies with established brand equity, startups must build everything from the ground up—but they also benefit from the freedom to start fresh without managing the complexity of multiple sub-brands and legacy products.

Many startups originate with a single product but then grow into offering multiple products and services under the original brand. For instance, Ring grew from a doorbell company to a leading home security brand. Amazon grew from an online bookstore to a huge collection of services that span shopping, entertainment, cloud storage, third-party marketplace, and much more. So, if you're building a startup with only one offering, be careful not to pigeonhole the brand or limit future growth. As we see from Ring's story, the long-term impact of early branding decisions can be profound. Make sure that the overarching company brand can expand to encompass potential future offerings.

Today the differences among brands, products, and features are blurring, especially in digital-only and service-oriented companies. For services such as Uber, Airbnb, and Spotify, a single name refers to the product, the brand, and the company. On the other hand, in some situations, unique features can become central to a brand's identity. For example, Snapchat's disappearing messages—known as "snaps"—became its major differentiator from other social platforms. These snaps almost became synonymous with the brand itself, reinforcing Snapchat's unique positioning in the social media space.

Whether you're branding a company, a product, or new features for an existing product, it's vital that they build upon each other to reinforce a single core identity. When a new feature is introduced, it should enhance the product's value, align with its existing branding, and support the company's overall promise. This integrated approach not only ensures a cohesive customer experience but also strengthens your long-term brand equity.

Repositioning and Rebranding

As businesses grow and evolve, repositioning and rebranding may become essential strategies for staying competitive, relevant, and unlocking new growth opportunities.

Before you start such a process, ask yourself what's driving a repositioning or rebranding. Does your business need an evolutionary, revolutionary, or disruptive change? Or maybe you don't really need a change at all?

HBO's rebrand journey illustrates how poor rebranding decisions can waste marketing investment, confuse customers, and dilute long-term brand equity. In 2020, HBO launched an independent streaming service that it called "HBO Max" to distinguish it from the core cable channels that were simply "HBO." In 2023, the company announced it would drop the word "HBO" from the streamer to call it "Max"—on the theory that it could broaden the streamer's appeal by highlighting content that didn't originate with HBO. But many users continued to call it HBO Max or even just HBO, which had been a premium brand for high-quality television since the 1970s.

In 2025, the company gave up and reverted the streaming service back to "HBO Max." It recognized that connecting it back to the roots of the HBO brand for distinctive, premium content was a huge positive, not a negative. As you can see from this costly and embarrassing misstep, branding decisions are high-stakes and should be carefully assessed before taking the plunge.

Clarifying your specific goals is critical to make sure these efforts will deliver meaningful and lasting impact. Your process may change depending on whether you need to revitalize a struggling business, respond to a merger or acquisition, execute crisis management, or adapt to broad market shifts.

Rebranding to Scale

Rebranding is never simply about changing a logo. It's about aligning the brand with a company's evolving vision, product offerings, and customer experiences to increase the value.

For example, in 2014, Airbnb realized that as their business grew globally, their community had outgrown the brand. To continue to scale and find new growth, Airbnb needed a brand to connect with people around the world. But their early brand was rooted in the technology platform, not the people.

So, they asked themselves, "What is Airbnb's mission? What is the big idea that truly defines Airbnb?"

Brian Chesky, CEO of Airbnb, explained, "For so long, people thought Airbnb was about renting houses. But really, we're about home. You see, a house is just a space, but a home is where you belong. And what makes this global community so special is that for the very first time, you can belong anywhere. That is the idea at the core of our company: belonging."[18]

It was this idea of belonging that crystallized Airbnb's mission and deepened its connections with their customers. As part of the transformation, Airbnb redesigned its brand identity, storefront, mobile interface, and customer experience, evolving from its old blue logo to a new coral-pink logo with a fresh design, which they called "Bélo"—short for belonging. It was designed to be "a symbol for people who want to welcome into their home new experiences, new cultures, and new conversations."[19] The goal of this brand refresh was to better reflect the company's mission and set it up for future growth.

This brand evolution wasn't merely an aesthetic change; Airbnb reimagined the entire product and user experience to bring this new identity to life. For example, Airbnb redesigned every page and touchpoint across both the website and mobile app, to elevate the experience and foster emotional connections with customers. From the look and feel to the user interface to immersive photography, customers can experience a more inviting Airbnb that celebrates community and local travel experiences. As part of the launch of this refresh, Airbnb also created a new commercial called "Don't just go there. Live there."[20]

This is an example of how rebranding created immense business value as it helped Airbnb establish its differentiation and distinguish its offerings from traditional travel and lodging options. This also laid the foundation and fueled the growth we saw in the next decade.

Rebranding to Maximize Value

Over the years, I have been involved in many branding and rebranding efforts. One thing I loved about working at Amazon was the opportunity to work with a range of strong brands, including Ring, Whole Foods, PillPack, and One Medical. I also learned so much about the nuances of making decisions while navigating a major brand like Amazon that had many complex and diversified businesses and product lines.

Debates about branding can get emotional. For example, when Amazon acquired the membership-based primary care service One Medical in 2023, we argued about how to position this new offering within Amazon and its existing health services. Should we lean more into the Amazon brand or the One Medical brand? How would either option affect the service's identity and customer perception?

Our team was divided nearly 50/50, with strong feelings on both sides about the long-term and short-term trade-offs. Amazon was a globally successful brand with strong equity, yet it was still building its reputation in the healthcare space. Adding the Amazon name to the One Medical brand would help that reputation. On the other hand, One Medical already had a good brand as a recognized innovator in primary care—why take a chance on changing that? The company ultimately chose to bring One Medical into the Amazon family and rebranded it as "Amazon One Medical."

A successful rebranding effort can solidify a company's or a product's relevancy to meet evolving customer expectations and continue to propel growth. A pitfall to avoid is to frame rebranding as a pure creative and design exercise, without considering the implications of product roadmap, customer experience, marketing investment, and, most importantly, its connection to customers. Think about the implications: When you rebrand, you may have to change your entire UX flow, product design, marketing assets, website, and app screens. It should never be done on a whim.

Repositioning and rebranding always requires significant time, energy, stakeholder alignment, and financial resources. You should be aware of the pros and cons to make a thoughtful decision that aligns with your business needs.

Forward-thinking investors increasingly evaluate brand strength as a key value driver when assessing potential acquisitions. This means strong brand identity doesn't just help customer acquisition—it becomes a critical factor in your company's valuation and exit opportunities.

Chapter Summary

Brand identity is a strategic long-term asset. Branding gives your product meaning and differentiation. Without it, your product becomes generic and doesn't stick in customers' minds.

Brand positioning, vision and story, and name and visual identity are fundamental strategic decisions that shape the overall brand foundation. These three foundational elements, coupled with tone, voice, and personality, work together as a system to evoke feelings and form customer connections.

A powerful way to bring your vision to life is through story. The human mind connects to stories better because we process information through stories, frames, and patterns. A strong brand narrative can help visualize what is possible and guide a potential product roadmap.

Naming is the most important branding decision you make. It's a one-way door decision that directly impacts product and business and requires a thoughtful and integrated approach to execute effectively.

Brand architecture creates cohesion between your brand, products, and features. This hierarchical structure simplifies your messaging to customers, makes marketing investments more efficient, and creates a unified experience that helps customers recognize and understand your offerings.

Repositioning and rebranding can be essential strategies to create value. Although it requires significant time, resources, and investment, strategic rebranding can transform a company and drive exponential growth.

Chapter 5

Integrate Your Brand into Your Product Experience

From Fragmented Customer Experiences to Products with Purpose and Personality

Establishing a unique brand identity is just the beginning—how do you bring that vision to life through the product you're building? Your product experience is your most powerful brand builder. When you embed brand thinking into it, you can transform your product into a meaningful, differentiated experience that resonates with customers.

This chapter explores how to integrate brand thinking into product design, ensuring every customer interaction reinforces your brand's identity, builds customer trust, and fosters lasting relationships.

The PillPack Story: Designing an Enduring Product

About 30 million US adults, nearly 10% of the population, take five or more prescription medications daily. Managing chronic conditions like diabetes, heart disease, and blood pressure while taking multiple medications every day is both complicated and challenging. Making multiple trips to the pharmacy, waiting in long lines, navigating insurance claims, organizing pills, and remembering to take medications on time while juggling doctor visits and maintaining their health—it's an overwhelming burden for so many.

This customer problem united TJ Parker and Elliot Cohen, the founders of PillPack. Parker was a pharmacist who had grown up working in his father's drugstore in New Hampshire, where he gained firsthand experience helping behind the counter, delivering meds to customers' homes, and managing various aspects of the business. He witnessed daily how customers struggled to manage their medications. Frustrated with the complexity and inefficiency of the process, Parker saw an opportunity to create a more consumer-friendly solution.

In 2012, Parker met Cohen, who was then a business student at MIT. Cohen's father had a lot of health issues. "When we were in the early days of conceptualizing PillPack, I was home in California when I interrupted my father sorting his pills into a traditional pill box—the 'day-of-the-week' kind. When my startled father turned around to see who had opened the door, he knocked over his pillbox, scattering medications everywhere. Seeing his frustration in that moment made me believe that there had to be a better way to manage medications. That was really the moment I committed to making PillPack a reality,"[1] Cohen recalled. They launched in 2013.

A Vision for a Better Pharmacy

This customer pain point was deeply personal to both founders. Their personal experiences not only inspired them to pursue a journey to build PillPack but also formed the foundation and vision of the company.

Healthcare is one of the least consumer-friendly industries in the United States, despite being essential to daily life. The traditional healthcare industry is often optimized for payers, providers, pharmacies, and manufacturers, not for customers. PillPack became the first e-commerce pharmacy and a pioneer in creating a consumer healthcare brand that redefined how people manage prescription medications. How did they create an enduring product that transformed and simplified the pharmacy experience?

In the early 2010s, modern consumer health service was almost unheard of. However, PillPack made a bold decision to focus solely on the end customers. They set out to solve a core customer problem: the complexity of managing medications for people with chronic conditions.

Design a Better System

Despite operating in highly regulated and complex areas, PillPack made designing a better system for customers its priority. Parker and Cohen were strong advocates of human-centered design. As they built their company, they knew they needed this to be a part of what they were building. They partnered with IDEO's Cambridge studio to work closely with the firm while incorporating human-centered design into their company's foundation.

Through deep collaboration, the team transformed its first prototype into the seamless service we know today. Medications are pre-sorted into personalized packets, labeled by date and time, and delivered every month. Best of all, the service itself is free. Customers only pay for their meds, with no additional service fees.

This is a life-changing experience for many. The pre-sorted packets format played a vital role in delivering a superior customer experience. No more sorting meds. No more chasing down refills. No more forgetting doses. Even over-the-counter medications, vitamins, and supplements can be included in the packets.

Customer service was another factor in improving the overall experience. For example, PillPack handled complex processes of working with payers and doctors in the background to remove friction. During severe weather events, such as hurricanes in Florida or blizzards in New Hampshire, PillPack's customer service and operations teams worked through the night to ensure customers received medications without delay.

Customers felt empowered and delighted. Susan, a PillPack user, shared her experience: "PillPack has been a godsend to me. I have a son who has Cerebral Palsy and is a quadriplegic. He depends on me solely for everything . . . Having PillPack means less pain for me and more time and energy towards my handicapped child. It is not that I just love PillPack. I need it."[2]

Integrated Product and Marketing

It's worth noting that the culture and team structure played a vital role in PillPack's success. The company organized its teams around the customer journey to create a seamless, end-to-end experience. For example, Colin Raney, who was then chief marketing officer, worked on product experience, customer acquisition, and brand development as a unified effort. This approach was fundamental to ensure consistency and cohesion.

As Raney explained,

We don't have a separation between marketing and product. PillPack is a service that works with customers over a long period of time, the marketing and the product very much intertwine because it's about telling a story about how we can help people stay healthier when they take their medications, and then it's about delivering on that promise in the product. When you do that, then people believe what you're about and they stick with your service.

Unlike most companies where product and UX design teams operate separately from brand and marketing teams, PillPack integrated these functions to create a cohesive experience. This alignment helped avoid a common pitfall in larger companies—fragmented efforts between marketing and product. When marketing makes bold promises that raise customer expectations beyond what the product can deliver, it ultimately erodes trust.

With PillPack, the experience was both the brand and the product, and it brought its vision to life across the product experience, marketing materials, and every consumer touchpoint, including website, dashboard, and a suite of physical products.[3] The integrated approach ensured that every customer interaction, from the first moment learning about the service to the sign-up process, onboarding, and daily usage, was smooth and seamless. More importantly, the unified approach ensured the product experience closely aligned with the brand's promise, earning customer trust.

As a result, PillPack earned customer love and industry recognition. *TIME* named PillPack one of the 25 best inventions of 2014. Cooper Hewitt, Smithsonian Design Museum[4] included the PillPack design in its permanent collection. In June 2018, Amazon acquired PillPack for $1 billion, and as of this writing, the service is integrated into Amazon Pharmacy's offerings.

We don't often associate simplicity and delight with healthcare, but PillPack showed that it was possible to build a consumer health brand that wins the hearts and minds of customers. More than 10 years after the launch, customers still talk about how the PillPack experience changed their lives.

As Parker reminisced about his early journey building PillPack, he said, "We made this decision that we were only going to focus on the end customer. In healthcare, you've got payers, you've got providers, you've got other constituents, you've got the end customer. Honestly, it's the reason that most healthcare isn't great, because almost no one is building for the customer that's using the service. We effectively put on blinders to everybody else, so everything we're doing is to make this easier for the customer."[5]

Obsessing over customers is easier said than done. This is one of the key reasons why PillPack became a pioneer in creating an enduring product and a consumer healthcare brand.

Why Brand Thinking Matters in Product Design

PillPack exemplifies the core principles of *Brand Power Built In*. By reimagining and redesigning the entire customer experience and consistently applying simplicity, user-friendliness, and care across every touchpoint of their product design and customer experience, PillPack created a product that not only solved customer problems but also earned lasting customer loyalty.

Brand thinking is integral to the success of designing an enduring product. Every day, customers experience a brand through its products. The product is the primary touchpoint for customers, directly shaping their perceptions, engagement, and loyalty. When you integrate brand thinking into a product, your product becomes not just functional but memorable and meaningful, you create a differentiated product with personality and purpose, and you create power to make customers choose you over the competition.

Unfortunately, many companies struggle to integrate brand and product to deliver a compelling customer experience. The brand and product developments often operate independently, and the misalignment, inconsistent messaging, and fragmented customer experiences sometimes lead to product launch failures and dilute the impact to drive customer appeal and growth effectively. This represents your lost opportunities, market share, and profitable margin. These missed connections with customers represent unrealized value that doesn't appear in your financial metrics. When you review monthly financials and wonder why product growth is lagging, this misalignment could be a critical factor. This is the hidden value we've been discussing.

How can you integrate brand foundations into product design? Let's explore five key lessons:

- Defining a shared vision and design principles
- Delivering on your promise with trust and transparency
- Weaving storytelling and emotional connections within the product
- Marketing and advertising as your product
- Connecting the dots to craft a seamless experience

From shared vision and design principles to product promise, from storytelling and emotional connections within the product to the entire end-to-end experiences, these lessons highlight critical aspects of product design and the journey of building memorable experiences.

As the legendary industrial designer Dieter Rams articulates in his timeless principles of good design:

"Good design makes a product useful."

"Good design is honest."

"Good design is long-lasting."[6]

When you design experiences with purpose, honesty, and longevity in mind, you create meaningful connections and enduring value, as we see from the success of PillPack, Apple, Airbnb, Rivian, and others.

Defining a Shared Vision and Design Principles

A shared vision and design principles are the foundation of a cohesive customer experience. Early work on positioning strategy and brand identity should inform product development and design and connect with customer-facing experiences. Yet, many companies fail to define a shared vision and design principles from the start. Skipping this step often results in products and experiences that feel disconnected from the brand they're building and the marketing campaigns that promote it.

A Shared Vision

A shared vision begins with foundational questions: What do you stand for? Who is the customer? What promise do you make? How do you differentiate from others? The same positioning strategy that guides brand development should extend to product vision, product requirements, and UX design, ensuring every aspect of the product and customer experience reflects and reinforces what the brand believes in and the value it delivers to customers.

From the beginning, PillPack founders Parker and Cohen had a shared vision for a simpler, better experience for customers. They didn't plan to make packets, but the packet design emerged as the best solution to holistically solve this customer problem. Customers want a subscription service so they don't have to make multiple phone calls and take multiple pharmacy trips. This is an example of how a clear vision and narrative can guide the product development process and deliver its promise to customers.

As RJ Scaringe, founder and CEO of Rivian, emphasized:

The best product has a clear vision for what it wants to be . . . That vision has to be communicated effectively across many groups. So there's a team of people doing power electrics, battery design, motor design, chassis design, interior design, exterior design, material design . . . and each of those teams need to be looking at it with the same mindset around how to make trade-offs between cost and features, between mass and cost, like all these trade-offs you have to make decisions where you can't make everything in everything.[7]

When designing products and services, trade-offs are inevitable. You have to balance speed, cost, and quality. It's impossible to get everything you want at the same time. You get to give up on something. But what do you give up? How do you make trade-offs? You have to align your decisions with your values. This alignment around a shared vision is vital to ensure that every decision across the entire organization follows the same principles and delivers a cohesive experience that customers can feel throughout the product.

Shared Design Principles

Product experience design is a direct reflection of the brand's identity. For a customer to have a memorable feeling about your offering, the tone, voice, and personality of your brand must come through in the product. This comes from leveraging shared tools and design principles to craft distinctive, memorable interactions across the entire customer journey.

There should be a consistent look and feel across all interfaces and user experiences that communicate the brand's core messages and values. It's like a translation process. You know what you stand for—your tone and voice, personality, imagery, color, and typography—and you make choices that reflect these elements in the product. This integration process is critical for extending these foundations into a digital product experience.

For instance, think about typography as a core element. We don't often think about font choices in our daily life, and it may sound like a small thing to any founder. The little voice in your head says: *It's just a font. Why does it matter so much?* Well, this foundational choice is the backbone of a visual system that communicates with your customers in both digital and physical experiences. From your website to your app to your packaging and marketing, typography directly impacts customer experience and guides customers through content, conveying your brand's distinctive style. So, typography isn't just an aesthetic choice; it's a significant brand and product decision. If you don't believe this, try putting the banner type of the *New York Times* into Helvetica, or the Google logo into Comic Sans.

Look for signature moments in the product where the brand should have a prominent presence and craft memorable interactions. The product experience should be reflective of that personality. Otherwise, you risk creating something generic without anything unique about your products and services. This integrated approach requires a design culture and mindset.

Delivering on Your Promise with Trust and Transparency

Having a vision is the starting point, but the execution of the vision and delivering it matters more. What the product delivers and how you communicate to customers need to be fully aligned.

A great tech product first needs to deliver functional value. Functionality is why customers buy a product. You can't connect with them without it.

Designing and delivering your product promise and value proposition is the most important element to build customer connection. Ground both the brand and the product in a promise that solves real pain points. For example, PillPack delivers on its promise to simplify medication management and make it easier for people with chronic conditions, through packets organized by day and time, delivered at no additional cost. Even though some other chain pharmacies launched similar packet services, they never solved the true customer problem. The real problem was never just about organizing and sorting pills. It was the overwhelming anxiety and emotional burden of managing conditions, from pharmacy visits to coordinating with insurances and doctors.

Without delivering on your promise, emotional connections fall flat because the product will lack the most essential foundation. Marketing and PR investments might as well go into the waste bin. Even if you leverage the best creative campaigns to drive people to your purchase funnel, you'll still find that customers won't buy your product. Then you'll have to go back to the start, focusing on improving your product. For a customer to care about your product, you first need to deliver value.

Avoiding Overpromising and Underdelivering

A common mistake companies make is overpromising and underdelivering. Setting realistic expectations is important to avoid disappointing customers. For any new products, what you deliver and how you communicate should be consistent. If the brand promises more than the product can deliver, it can break customer trust—the foundation of lasting customer relationships. And without trust, even the most innovative product can fail to gain traction. Balancing trade-offs to build trust is especially important when your brand needs to resonate emotionally, even during friction points when customers may encounter difficulties while engaging with your service.

When designing an experience, it's essential to identify moments where trust could be broken, where the product may not be working as well as it could or should. How do you maintain trust during those moments? When errors or issues arise, how can you reestablish trust? You must think about both "happy" and "unhappy" paths and what the customer experiences look like.

True customer obsession sometimes requires financial sacrifice in the short term. For example, PillPack had a full in-house customer service team, which required investment and manual hours of customer service agents, but it was worth the investment to improve the customer experiences. It was a deliberate decision the company made to prioritize experience and satisfaction over cost.

These decisions are not always straightforward. Companies striving for profitability may need to make difficult choices, such as cutting services or scaling back investments in customer experience. In these scenarios, brands can manage expectations thoughtfully and maintain trust by being transparent about trade-offs.

Today's customers have high expectations beyond the actual product or service. They expect companies to act as a positive force for the world. Trust isn't limited to quality and reliability; it's also tied to broader responsibilities for humanity. For example, when designing a digital experience, privacy must be top of mind. When designing packaging, the impact on the environment must be considered. When developing an AI service, ensuring ethical uses and safety is a challenge. In all of these and other situations, thoughtful design is a great way to show respect for your customers.

Weaving Storytelling and Emotional Connections Within the Product

Companies often spend significant effort and investment to tell stories about the product and promote it through beautifully crafted advertising campaigns, yet they don't put the same level of rigor, thought, or investment into in-product storytelling. It can be a huge missed opportunity if you don't prioritize the product experience ahead of anything else.

In-Product Narrative and Storytelling

UX is not about aesthetics; it communicates the heart and soul of a brand through customers' interaction and experience with the product across digital and physical mediums. The integration of branding and UX design is critical as it directly impacts customer perceptions, engagement, and business results. UX wireframes, visual elements such as font and color palette, and messaging need to work together to make people feel something and inspire customers to take action.

I once spoke to a product design leader at a major tech company that made a difficult decision to lay off all of its UX writers; the product managers had to write in-product copy instead. The assumption was that product writing shouldn't be that difficult. Product managers know the features, so they should be capable of explaining those features to customers. It's just a few words, right?

A few months after the layoff, almost all of the product managers admitted that they actually couldn't write effective copy. During the team's senior leadership reviews, as much as 80% of the negative product design feedback from leadership was about confusing or unclear text. The in-product communications felt like 10 different people had worked on them. They were too technical, wordy, and disjointed, without evoking any feelings for the product. This product design leader was devastated about the loss of his team, but also frustrated because senior leadership should have known that UX writing was important. Instead, they made the mistake of trying to do it on the cheap. "They want it, but they don't want to pay for it."

This is a fairly common experience that product leaders face at many companies. Like systems and technologies, clarity, tone, and voice of messaging can offer a huge opportunity to improve. Can you invest more in in-product communication and storytelling? It might pay off more than your investments in advertising.

Speaking of advertising, I've also seen leaders fail to invest in good creative work because they think, *I can do the copywriting myself. It's not that hard.* If you're serious about building great products and brands, you'll need care and craft to make *every* message in *every* medium more engaging and compelling.

Great design is the added effort that can't be captured in product requirements. Consistent and thoughtful UX writing is crucial for delivering your product's promise, establishing connections with customers, and evoking feelings. Conversely, inconsistent and poor experiences can erode product quality and weaken customer connections and trust, impacting business growth in the long term.

Emotional Connection and Delight

Creating delight in a product is an art. Your product's personality needs to come to life to make people feel something. How can customers *feel* your brand when they use the product? How can engineers *feel* the brand during the design process? These feelings should be the same for both the end customer and the product developer.

Your product can communicate and talk to people through its promise and storytelling across the user experience. People like to interact with UX/UI, so try to embed the brand's story and bring it to life in a more emotional and memorable way.

I'll never forget when Apple's product design leader Alan Dye came to TBWA\Media Arts Lab to talk about how he and his team designed the interface for the first-generation Apple Watch in 2015. I was intrigued by the detail that users could customize their Apple Watch interface, from Mickey Mouse to a flower blooming to a moving jellyfish. I learned that the flower blooming motion wasn't computer generated; the flower had been filmed for hundreds of hours to capture its motion.

Dye said, "I think the longest one took us 285 hours, and over 24,000 shots." His favorite was the Earth interface. "We worked really hard with our engineering team to make sure the path you take from your actual position on the Earth to where the moon is and seeing its phase, is true to the actual position of the Earth relative to the moon."[8] This small delight makes people's hearts sing, even though it's not part of the watch's core functional features (like its Activity and Heart Rate apps). It's the combination of functional features like health applications along with the subtle delight in the experience that makes Apple Watch different and unique.

Apple's focus on how a product makes people feel reflects the importance of emotional resonance. As the "Designed by Apple in California" campaign states: "This is what matters. The experience of a product. How it makes someone feel."

This approach reminds us that emotional moments and delight are designed and engineered *inside* the product. This is a core message of the *Brand Power Built In* approach.

Marketing and Advertising as Your Product

Another major lesson I learned from Apple in my early career is that every marketing and advertising campaign should be treated as a product. That's why Steve Jobs insisted on the same high standards for marketing and advertising campaigns as he did for new product launches.

Think about your marketing campaigns and content as part of the customer journey and connect directly to the product experience. This helps broaden your understanding of what constitutes a product in the eyes of customers. Product is not just limited to what it does functionally;

it's also about how customers learn about it and discover it. Customers don't separate these moments and touchpoints, as they think it's all about your product.

When you start with the customer journey, it forces you to think about the totality of the experience. From marketing and advertising campaigns to landing page content and the product experience, mapping the customer journey provides a holistic view of how each element of product development and marketing connects to and amplifies every other element.

There are three phases of the customer journey, and each plays a role in shaping customer perceptions and experiences. They're the embodiment of your brand.

- **Awareness and Discovery:** Advertising, marketing, PR, search, company blog, and partnership efforts that raise awareness and drive discovery of your offering.
- **Acquisition and Engagement:** Sales channels, website, packaging, app, buying experience, and onboarding that drive conversion and initial engagement with your product.
- **Retention and Loyalty:** Product experience and updates, customer service, email, notifications, community, and repeat purchase activities that retain customers and build loyalty.

Solve Product Design Challenges Through Marketing

When you look at the entire customer journey, you have a holistic view of how product experience, marketing and advertising, customer service, and other functions all work together from a customer's perspective. This approach allows teams to break out of traditional silos and think more creatively about cross-functional solutions.

For example, marketing can play a crucial role in product design challenges. When we worked on Amazon Pharmacy, we needed to build

Customer Journey Touchpoints

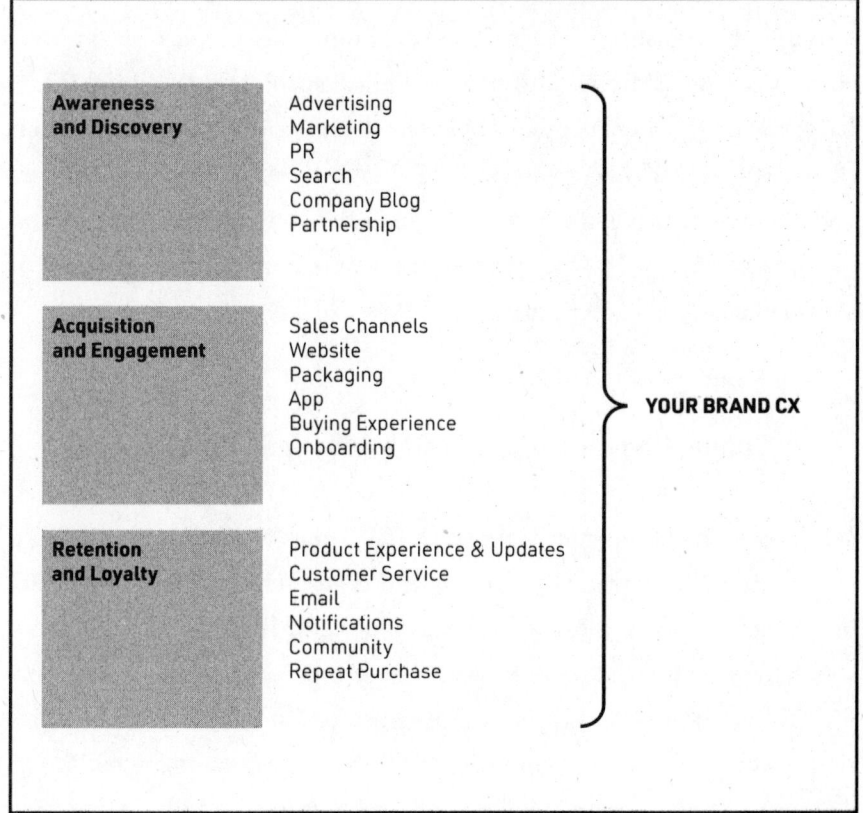

Awareness and Discovery	Advertising Marketing PR Search Company Blog Partnership	
Acquisition and Engagement	Sales Channels Website Packaging App Buying Experience Onboarding	YOUR BRAND CX
Retention and Loyalty	Product Experience & Updates Customer Service Email Notifications Community Repeat Purchase	

trust with potential customers and help them understand how the service works. Customers are used to talking to their pharmacists and picking up their medications directly from the counter. How do you build trust for a new online pharmacy that requires customers to change their behaviors? How can you remove potential concerns and help customers understand how Amazon will address their medication needs? Product teams can't incorporate *all* educational content into the UX flow, as that would increase complexity. When you add more text and steps to the sign-up flow, you could lose customers in the process. The added work would also require far more engineering resources.

As an alternative solution, our marketing team produced an educational video that showcased the end-to-end customer experience through the journey of a single medication. This medication journey video[9]—a much simpler solution—was placed prominently on the "How It Works" page, where people could easily learn more about how Amazon Pharmacy works. This cross-team collaboration effort helped customers understand the product and address potential concerns about switching to a new pharmacy. It was a great example of making marketing a part of a unified customer experience.

Connecting the Dots to Craft a Seamless Experience

Building a great customer experience (CX) is at the heart of creating a strong brand. Your product's end-to-end customer experience is a direct interaction between a brand and a product and its customers and how you deliver your brand promise. Every customer interaction impacts how they think about you. When you design a unique, differentiated experience, you can charge a premium price for it. Customers are willing to pay for great services and experiences, and competitors can't easily copy your customer experience.

Don't Fragment Your Customer Experiences

When you build a new product from the ground up, teams operating in silos can be risky. If each team works on their own deliverables and pursues their own success metrics, the result can often be fragmentation and disjointed experiences for customers.

For example, you might miss an opportunity to connect the dots from in-product storytelling to marketing and sales and the entire end-to-end customer experience. The product team often focuses on one message, but when the product gets passed to marketing, the marketing team comes up with a different narrative. Then, when an advertising agency is brought in to help with the launch, they may come up with yet another narrative. Instead of fragmenting your customer experiences this way, it's essential to align all the elements and craft a seamless experience.

Airbnb faced a fragmentation challenge when the company scaled rapidly around 2019, before COVID hit. Each internal team had its own product roadmap, and other teams and regions had sub-roadmaps, creating dependency and complexity. At one point, even CEO Brian Chesky didn't know what each team was working on, because there were too many roadmaps and mini roadmaps.

Chesky recalled: "We had like 10 different divisions, each with like 10 different subdivisions. We were very much run by product managers. We had a plethora of A/B experiments. The more people we added, the more projects we pursued, the less our app changed and the more the cost went up. I didn't know what to do."[10] To simplify and focus, he asked everyone to put everything on one roadmap. After reviewing the work, he only focused on 10% of the product plan and reduced the other 90% to the most basic options. "Simplify, simplify, and simplify" was the new ethos. They began to focus only on shipping features that they felt most proud of.

Another problem Airbnb faced when they scaled is that the design team became a service organization. This made it difficult to design a more cohesive and holistic experience. To reduce organizational fragmentation, Airbnb streamlined the company strategy, organized by functions, and embraced a leaner operating model. For example, the company combined UX writing[11] with marketing writing, ensuring a consistent voice across the app, email, product communication, and advertising to deliver a better customer experience. This change was strategic to the business as it helped the company scale more efficiently.

End-to-End CX Walk-Through

Steve Jobs was a master at connecting the dots across product design, marketing, and advertising. He used to review all global marketing and advertising campaigns every Wednesday. But most companies are not as centralized as Apple and can't rely on one person to review all aspects of the customer experiences holistically. You will have to connect the dots through team-driven mechanisms.

I learned the power of repeatable mechanisms to drive better results at Amazon. Good intentions alone don't guarantee results—you need good mechanisms to make anything happen consistently. The book *Atomic Habits* also makes a similar point: Lasting change comes from building the right habits and creating systems for personal transformations. I believe the same principle applies to organizations. After all, organizations are made up of people who drive change together.

For example, doing an end-to-end CX walk-through is a great way to understand the entire customer journey and how different parts of that journey fit together. The walk-through usually starts with how a customer discovers your new product, generally through marketing or PR efforts. It then showcases how a customer engages with your website—whether they purchase, sign up, or leave—and explores their subsequent experiences based on what the customer chooses to do. A CX walk-through is often led by the product design team, but it should be a shared responsibility across all departments that contribute to creating and reviewing the end-to-end experience.

When you bring together product, brand, marketing, customer support, and others and rally around the entire customer experience, you break down organizational silos and ensure every part of your business works together to deliver the best possible experience. Crafting a seamless experience requires intentional choices at every stage, from discovery to engagement to repeatable usage. Every stage must be consistent with what the brand believes in.

Chapter Summary

Defining a shared vision and design principles. This shared understanding of how brand foundations translate into product decisions ensures every aspect of the experience reflects what your brand stands for. When teams align around these principles, they make consistent decisions that build stronger customer connections and differentiate your product in meaningful ways.

Delivering on your promise with trust and transparency. When you fall short, be transparent to set the right expectations and build trust. Setting realistic expectations is important to avoid disappointing customers.

Weaving storytelling and emotional connections within the product. Product experience is the most powerful brand-building tool. In-product narrative and storytelling is often an undervalued and underinvested area to build emotional connections and great experiences.

Marketing and advertising as your product. Broaden your understanding of what constitutes a product in the eyes of customers. Product is not just limited to what it does functionally; it's also about how customers learn about it and discover it. This integrated approach ensures that you create a seamless experience across all touchpoints.

Connecting the dots to craft a seamless experience. When you bring together product, marketing, and customer support to review the end-to-end customer experience, you break down organizational silos and ensure every part of your business works together to create a powerful, memorable experience.

PART III

Launch, Scale, and Iterate

We began the journey by laying the foundation for building a better brand and product experience together. When you create a differentiated product with personality and purpose, you establish the groundwork for go-to-market success.

But the *Brand Power Built In* approach doesn't stop at the product experience—it extends into every phase of the customer lifecycle. From go-to-market strategy to customer acquisition and retention, and through continuous iteration and measurement, the goal is to leverage what you're already doing and ensure that every effort contributes to winning customers and driving sustainable growth.

In this part, we'll continue the journey by focusing on the next three building blocks:

1. Launch with a Sequenced Go-to-Market Strategy
2. Scale Your Brand to Accelerate Acquisition and Retention
3. Iterate After Measuring Impact

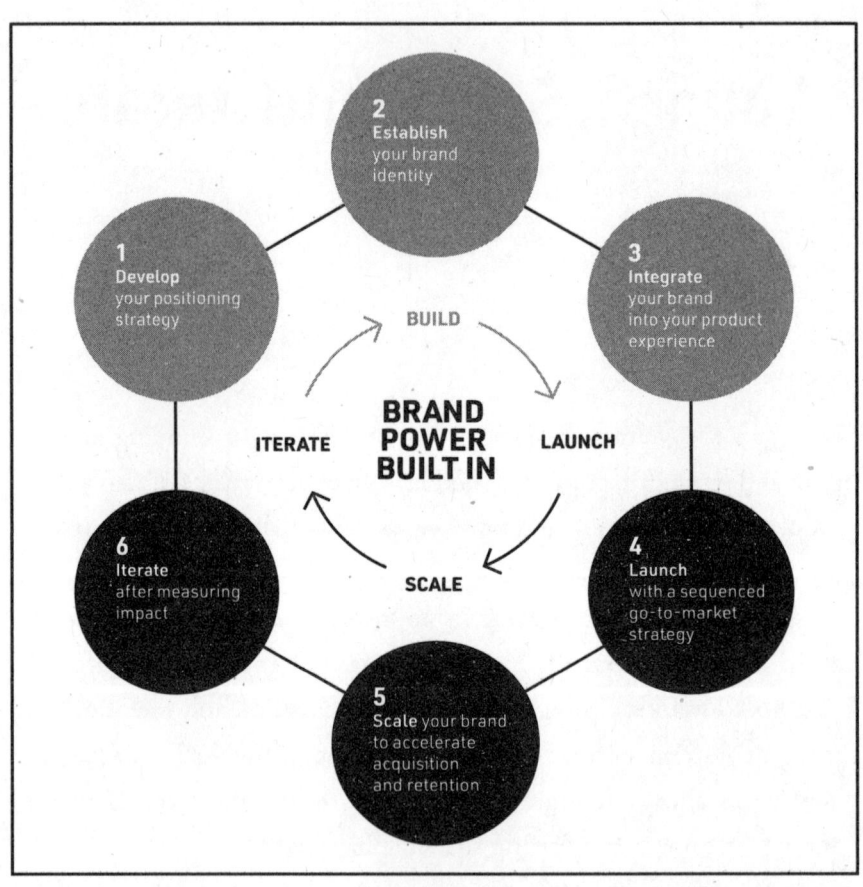

Chapter 6

Launch with a Sequenced Go-to-Market Strategy

From Pushing Your Product to Pulling Customers In

Too often, companies push products out to customers, hoping for adoption. But the most successful launches create traction and naturally pull customers in. When you build in brand power from the start, you set the foundation for a stronger launch that will drive faster market adoption and lower customer acquisition costs.

This chapter shows you how to build that magnetic pull, from laying your foundation before launch to achieving product-market fit to sequencing your customer acquisition strategy.

Notion's Pivot to Redefine Workspace Collaboration

How did an app-building software tool that almost went out of business pivot to become an all-in-one productivity tool used and trusted by over

60% of Fortune 500 companies? Notion offers features like shared documents, note-taking, and project management, plus AI that takes notes, searches apps, and builds workflows, competing with Google Workspace and Microsoft 365. Since its first release in 2016, Notion has grown from a scrappy product to a profitable business with more than 100 million[1] users and valued at $10 billion.[2]

However, Notion's journey wasn't smooth. Like many startups, it faced near-failure before finding a compelling use case that led to its current success.

In the early days, cofounders Ivan Zhao and Simon Last had an idea to build software that would help people who didn't know coding create their own apps, websites, and tools. After working on the project for almost three years, they realized that the product they were building wasn't on the right path. In the summer of 2015, Notion ran out of money and almost had to shut down. It turned out that there wasn't a real customer need for app-building software. As Zhao explained, "We focused too much on what we wanted to bring to the world. We needed to pay attention to what the world wanted from us."[3]

It was a painful lesson many startups face.

Disappointed about the outcome, Zhao and Last scrapped the project. With a $150,000 emergency loan from Zhao's mother, they moved from San Francisco to Kyoto, Japan, a less expensive place to start all over again.

The Pivot

After working 18 hours a day, Zhao and Last pivoted the product to focus on work collaboration via documents, notes, wikis, and project management, with a goal of making it radically easier for people to do their jobs. The new strategy felt more promising because productivity is a far more universal need than building apps and websites.

Once they identified this gap in existing workspace tools, they laser focused on building an all-in-one productivity tool that would combine ease of use, beautiful design, and customization to deliver a more user-friendly experience. After building and rebuilding Notion 1.0 four times, they released the product as free software in March 2016. It gained immediate traction among the tech community and earned the No.1 Product of the Day on Product Hunt, a community website where people share and discover new tech products and services.

The release of Notion 2.0 in March 2018 was even bigger. The success was evident in its rapid user adoption, indicating a strong product-market fit. It earned the No.1 Product of the Day, Week, and Month on Product Hunt for a second time. The product even got a great review from the *Wall Street Journal*, which called it "The Only App You Need for Work-Life Productivity."[4]

In September 2019, Notion reached one million users worldwide. According to *Forbes*'s estimate, Notion made $250 million[5] in revenue in 2023. Teenagers now use Notion to manage their hobbies and collaborate on school projects. Employees at OpenAI, Toyota, and 50% of the most recent Y Combinator class use Notion to manage products and share notes.

Expansion Powered by Community

How did Notion achieve this go-to-market success? The biggest factor was encouraging community engagement to drive word-of-mouth advocacy. As of 2021, 90% of Notion's growth was organic,[6] driven by direct traffic.

Notion's involvement in community support and direct communication with users on Twitter created a connection and cultivated its fan base. Notion also tapped into power users, gave early access to new features, and invited them to provide input and feedback. Its power users then advocated for the brand and supported its expansion on a grassroots level.

This community-driven approach also allowed the company to gain insights into trends and customer feedback to improve the product. For example, Zhao gets notifications about every customer service problem on his phone. Customer service shouldn't just be some department that fulfills a task. According to Zhao's interview with *Entrepreneur*,[7] it must be part of a system, integrated into the whole like a beetle's mouth is connected to its guts, so that every part of the system can learn from customers, build based on their feedback, and make the customer happy so they return.

Zhao envisions Notion as the LEGO of productivity tools and wants to make it an "AI-everything app" for the office. "In the next 5 to 10 years, Notion could be the front-end infrastructure for the world. Notion takes care of search, notifications, permissions. Just dream of a piece of software, and you should just be able to build that using Notion."[8]

Notion's journey from a struggling app-building tool to a $10 billion company shows that growth is rarely linear. Their success wasn't built on a broad launch with high marketing spend. It was the result of a sequenced approach that first established itself in the tech community through word of mouth, then gradually expanded into the broader business world, eventually becoming the productivity platform we know today.

Why Sequenced Growth Matters

A sequenced growth approach is critical to tech product launches. Go-to-market efforts often fail without a strong foundation and proper sequencing. Almost no new product is an overnight success. For most product launches, the journey looks a lot like the following graph. The building phase takes a few years. If the launch doesn't gain momentum, you need to pivot and redesign to improve the product experience until you find product-market fit.

The Path to Product-Market Fit

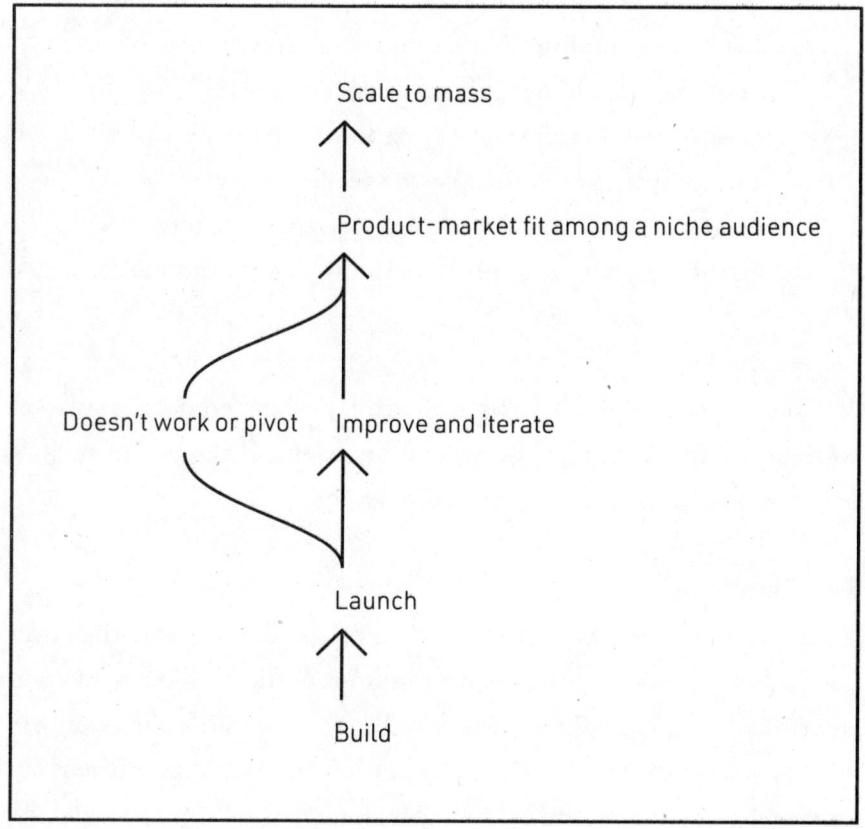

When you launch something new, there are many unknowns and uncertainties. Your product might face performance issues, your operational capacity may be unprepared, or your resources and budget might be insufficient to support growth. A minimal viable product or first-generation product will likely appeal only to early adopters who are open to trying new things that aren't perfect yet. This is different from product launches like the next generation of iPhone, which has already been tested and proven in the market for mainstream customers.

Watch out for these common pitfalls that kill promising product launches:

- **Misalignment with target audience:** Products are not grounded in real customer needs and lack a clear market fit.
- **Lack of marketing:** Without marketing, even the best products can fail to gain market traction.
- **Overly broad targeting:** Trying to appeal to everyone at launch instead of focusing on a specific audience.
- **Premature scaling:** Expanding too quickly before establishing a unique position in the market and validating product-market fit.

These common pitfalls often stem from pushing products to market without careful planning and preparation. Many of these launch challenges can be avoided during the building phase.

Pull, Don't Push

Today, customers are bombarded with more product choices than ever before. Simply pushing your product in front of them via big marketing investments is no guarantee of success. But a sequenced launch can help solidify your market position over time, while making more efficient use of your time, resources, and budget. This strategy also helps you avoid the risks of overpromising and underdelivering, especially if the product isn't ready to meet expectations right out of the gate.

I learned this lesson in a hard way during my first year working on Amazon Key In-Garage Delivery. We faced an awareness challenge: No one had heard of us. To accelerate awareness, we came up with an ambitious plan to go big and blast customers with a fully integrated advertising campaign. We proposed a multimillion-dollar marketing plan with the assumption that more awareness would make customer acquisition easier.

But when we presented it to the senior vice president, he asked me a tough question: "What data points give you confidence that this will work?"

At that moment, I realized we had completely missed the mark. The meeting went poorly. Our proposal got rejected. Yes, awareness was important, but In-Garage Delivery wasn't a mass-marketable product. The service was only available in select markets, customers had to have connected garages, they had to shop a lot on Amazon to want to get their goods delivered inside the garage, and they had to be Prime members too. All of these factors would limit our chance at rapid growth. Simply throwing a big budget at a mass marketing campaign wouldn't solve the problem.

That moment taught me an important lesson: It's better to start small, test an approach, and prove ROI before scaling. When you define your go-to-market scope and consider when and how to scale your marketing activities, you need to assess criteria like your business priority, product readiness, target audience, budget, and acquisition economics, since each variable can influence your campaign's overall effectiveness.

As of this writing in early 2025, Apple Intelligence is experiencing the backlash of overpromising and underdelivering. The Apple Intelligence launch campaigns that began in 2024[9] went big. At every Apple Store, you saw promotional materials on the wall marketing the iPhone 16 as being "Built for Apple Intelligence." Apple's partners, including AT&T, heavily supported the advertising and marketing campaign. But the ongoing delays of Apple Intelligence features triggered a federal lawsuit[10] for misleading consumers about the product's actual utility and performance.

Having launched new brands, products, and services across diverse categories and global markets, I've seen both remarkable successes and disappointing failures. Sequenced growth is critical for tech products, particularly given their technical complexity and performance, and even industry leaders can stumble when they fail to properly sequence their launch strategy.

The Danger of Expanding Too Fast

Early electric scooter companies like Bird and Lime illustrate the danger of rapid expansion. They flooded cities with electric scooters, and the hyper-growth strategies resulted in operational chaos, regulatory issues, and financial losses. In 2023, Bird filed for Chapter 11 bankruptcy protection[11] when the service was available in 350 cities around the world. The company faced severe financial losses and legal troubles, including being delisted from the New York Stock Exchange for failing to maintain the required $15 million market capitalization for 30 consecutive days. (A year later, Bird announced that it became part of a newly organized private parent company, Third Lane Mobility.[12])

In contrast, Waymo One, the world's first autonomous ride-hailing service, took a measured approach. They spent years testing in Phoenix, Arizona, before expanding to other cities such as San Francisco and Los Angeles. Along the way, they proved their technology in each market, built public trust, and optimized the service and customer experience.

When Waymo opened its service to all Los Angeles residents in November 2024, nearly 300,000 people had already joined the waitlist. This sequenced geographical rollout allowed Waymo to leverage local adoption to overcome the significant trust barrier inherent in self-driving vehicles. Waymo's initial expansion in LA generated high customer satisfaction: "98% are satisfied with our service and 96% find it useful."[13] This customer feedback, combined with organic word of mouth and optimized ride experience, was vital to create momentum and gain traction for the new service.

Anecdotally, a friend of mine tried Waymo because other people told her it was excellent. When she and her husband both tried it, they really liked the experience too. They got to ride in a brand-new Jaguar EV, with the ability to play their own music and no worries about whether the driver was tired or distracted.

Waymo's sequenced approach was deliberate, giving the company enough time and opportunity to perfect these and other experience details. When people have a great experience, they will tell their friends and families about it. Word of mouth is especially powerful in helping people alleviate concerns about something potentially scary, like a self-driving car.

The contrast between Bird and Waymo offers a powerful lesson in sequencing: Instead of pushing a product widely on day one and hoping for mass adoption, successful companies build up traction and credibility with a core audience first.

So how do you build that customer traction and credibility? Whether you're introducing an innovation, entering new markets with a unique offering, or launching a repositioning and rebranding effort, your success depends on the foundation you build before launch.

Laying Your Foundation Pre-Launch

Most go-to-market strategies focus on the moment of launch—channels, campaigns, growth tactics, and launch readiness. But the most successful launches begin much earlier, during the building phase. When you build in brand power in the early stage, you lay the groundwork and set the stage for go-to-market success. It starts with three critical foundation elements.

First, your positioning strategy is the backbone that informs product development, brand development, and go-to-market (GTM) efforts. Positioning helps you define your product's place in the market and align with customer needs.

Second, your brand story gives your product meaning and differentiation. When you integrate your brand into your product experience and the end-to-end customer touchpoints, you create powerful, memorable customer experiences that form stronger connections.

Third, early development in customer segmentation, value propositions, and messaging can guide the GTM approach and provide insights into the target market, pricing, and channel selection. This integration across product, brand, and go-to-market ensures that all aspects of the launch—from positioning to channel selection to customer experience—are aligned, helping attract the right customers and establishing the product in the market.

This early foundation sets the stage for product-market fit and creates momentum for faster market adoption and lower customer acquisition costs. Without a proper foundation early on, pitfalls often emerge during the go-to-market phase. When products fail to gain traction and you ask why the business isn't growing, you may realize you skipped foundational work that could have avoided later challenges.

For example, the initial idea for Notion was not grounded in real customer needs. The founders didn't build enough foundation in defining their positioning, their customer profile, or their value propositions. When they went to market, they learned that the target audience for an app building tool was too small to turn the idea into a real product. But after learning from those early mistakes, the founders successfully pivoted to compelling use cases and focused the product on work collaboration and productivity.

Finding Your Product-Market Fit

Achieving product-market fit is a major milestone that validates the viability of a new product and business and gives you confidence to scale. Laying your foundation also helps you assess product-market fit earlier, rather than waiting until after launch to figure it out. The positioning strategy defines your product's value proposition, customer segments, and competitive advantage. When you launch, you can test these hypotheses through real customer feedback and market reaction. How will you know when you've truly found product-market fit? What can you do if your launch isn't gaining the traction you expected?

It's important to understand that there's no specific moment that definitively signals, "Great, now we've achieved product-market fit." It emerges gradually through ongoing positive feedback from customers. You're looking for a trend toward customers who care about your product, who are willing to pay for it, and who recommend it to others.

Focusing on Your Niche First

Whenever you launch something new, be mindful of going too big too fast. Focus on early customers who love your product and build momentum from there.

Matt Ridley, the author of *How Innovation Works*[14] and an expert on the history of innovation, reveals a surprising pattern: Most innovation is gradual and incremental. You rarely find sudden breakthroughs; it's almost always "a series of small incremental steps." This insight speaks to why big, splashy launches don't always translate to big successes.

Most startups fail in the early phases in the first few years. The main reason is that many make the mistake of doing too many things too quickly and not focusing on perfecting and improving the product experience before investing in customer acquisition. A friend of mine learned this valuable lesson when she shut down her startup after pouring five years of heart and soul into building a social e-commerce platform. To avoid that fate, try to leverage insights from your first 100 or 1,000 customers to improve the product and ensure the experience is optimal before considering doing more.

For example, as we discussed in chapter 3, when Rivian entered the EV market, it first focused on the high-end pickup truck and SUV market with the R1T and R1S. It didn't go after everyone. Its clear brand positioning, focused on the luxury customer segment, was vital for its early go-to-market success. Rivian didn't offer a mass-marketable product until the announcement of two mid-priced product lines, with R2 (starting around $45,000) scheduled for release in 2026 and R3 (starting around $37,000) in 2027. From the first Rivian car debut at the Los Angeles Auto Show in 2018 to the mass product line announcement for R2 and R3 in 2024, it took nearly six years to grow from a niche to a more mainstream brand. Rivian's sequenced go-to-market approach shows how you can start with a focused market before expanding to a broader audience.

Iterating Through Customer Feedback

Once the new product is out in the world, it's important to gather feedback and data to ensure that it meets customer expectations, and to evaluate the effectiveness of your go-to-market approach.

To gather and analyze customer feedback, consider user surveys and interviews along with analyzing usage data and other customer metrics, if possible. Such feedback is valuable to validate your positioning (targeting audience, value proposition, and the appeal of your product) and assess what works and what doesn't. These early signals can help you improve the product experience as well as adjust the GTM strategy.

Some key questions to ask:

- **Product adoption:** How do the market and customers react to your new product?
- **Target audience:** Does your product gain traction among your target audience?
- **Customer experience:** What does your target audience value most about your product?

In the early days, Notion prioritized developing core product features centered around notes and document creation. Their strategic decision was to perfect these fundamentals before adding more features. If the core functions worked very well for early adopters in the tech community, they would drive word of mouth. So Notion focused on serving startups in San Francisco before expanding to other parts of the country. Early users loved the new experience and quickly adopted the service. They appreciated customization, simplicity, and intuitive user interface for workspace collaboration. It took three years for Notion 1.0 and then Notion 2.0 to reach one million users.

Notion's go-to-market strategy leveraged a universal need: documents and notes. With an estimated one billion knowledge workers worldwide requiring these basic productivity tools every day, the company strategically positioned notes and document creation as their entry-point features. Notion leveraged these universal features to raise top-funnel awareness, attracted customers to try the service, and encouraged them to explore its more advanced features. Gradually, users came to discover its project management and collaboration capabilities. This approach worked exceptionally well. Almost half[15] of Notion's B2B enterprise customers discovered Notion from their personal use experiences.

When you pursue product-market fit, you'll need patience, flexibility, and a strong focus on customer feedback. By building a strong foundation, launching to a niche audience, and constantly iterating based on feedback, you can significantly increase your chance of creating a product that truly resonates with a wider target audience.

Sequencing Your Customer Acquisition Strategy

Similar to how you systematically find your product-market fit, you need to sequence your customer acquisition strategy. Targeting, channels, and economics will all shape your acquisition results. Let's take a closer look at each.

Effective Acquisition Starts with Targeting

A few years ago, I met a chief marketing officer at a growing consumer healthcare brand whose team managed a $200 million annual paid media budget. When I asked him, "Who is your customer?" he replied, "We're for everyone." This response signaled a lack of precise targeting. On a deeper level, it was a red flag that his team didn't understand their core customers and therefore their marketing investment might not work as intended.

Effective customer acquisition requires a deep understanding of your potential target audience, their purchase journey, and the best channels to reach them, whether through digital marketing, social media, retail partnerships, or direct sales. You can build on the customer segmentation work done during your positioning phase. Identify a subset of target audiences and develop tailored channels and campaign content to reach and convert them.

For new products, it's important to segment your potential customer base and identify high-intent customers versus low-intent customers. So, it's strategic to prioritize high-intent customers first to increase effectiveness and efficiency.

For example, Notion's early growth and expansion largely focused on individuals, tech and creative communities, and small teams, to establish its reputation as a better productivity tool. As Notion's customer base grew to include more businesses, the company then hired its sales team and provided more enterprise support more broadly.

Prioritizing Owned and Earned Channels

Today, you have more ways than ever to reach customers, yet getting their attention has never been harder. You must innovate in how you connect with customers. Customer acquisition comes from three main sources:

- **Direct organic growth:** Driven by PR, word of mouth, organic search and discovery, and brand strength

- **Direct paid growth:** Driven by paid marketing activation such as paid search, social, display, and online video
- **Sales and distribution channels:** Driven by retail partnerships, wholesale, and partner activation

How you leverage these levers is largely dependent on your business goals and priorities. However, you need to first focus on building and delivering a great product and customer experience. Before investing in paid acquisition, prioritize optimizing owned channels—your website, app, email, or organic social—to ensure you don't have a leaky customer acquisition funnel. Then you can focus on targeted audiences to drive demand before expanding to mass audiences.

Large-scale sales and distribution investments should wait until you prove product-market fit. Experiment with paid marketing channels to gather learnings, but wait to expand reach until your product and customer experience is optimal. Otherwise, you risk disappointing potential customers who arrive at your website or download your app only to find the experience doesn't meet their expectations.

Improving Acquisition Economics

If you lead marketing, improving acquisition economics is probably one of your most difficult tasks. Controlling acquisition costs is critical to building a sustainable business.

I experienced this challenge firsthand while working with a particularly frugal founder. He built his company in a garage with cheap chairs and tables, only spending on the most essential things. Every time I put together a marketing plan, I can still hear his voice in my head, questioning my budget and approach:

"Why do we need to spend money on this?"

"What's our ROI?"

"Is this a smart investment?"

That's the reality of a startup and of building a sustainable business, even as you scale. You have to be frugal about marketing spend, push for the highest possible ROI, and constantly make trade-offs while balancing savings and smart investments.

This work can't be done alone. You must work with your finance team early to model customer acquisition costs and set up an effective measurement framework. Invite your finance partner into the marketing process, build relationships, help stakeholders understand the business challenges, and be realistic about growth expectations. It's critical to set guardrails and run experiments to understand and prove your financial model as you grow.

If the business is not growing as fast as you expected, investigate which part of the customer journey is the main problem:

- Is it a product challenge?
- Is it a price challenge?
- Are you marketing to the wrong audience?
- Do you have a leaky conversion funnel?

When you uncover the key obstacles to effectively acquire customers, it almost always comes down to one of these: a product offering that isn't compelling enough, poor targeting, ineffective channels, or lack of awareness about the value proposition. Once you identify the root cause, you can fix what's actually not working.

Pulling Customers In Through Effective Storytelling

When launching a product, you can't underestimate the power of storytelling in capturing people's hearts and minds. Effective storytelling isn't just a creative exercise; it's a strategic tool to pull customers in. At launch, people form their first impressions of your product and your company.

Today, customers discover and engage with your product across many touchpoints. Whether through a TikTok video, a homepage visit, or a press article, your story must come through cohesively across these moments to shape perceptions, drive adoption, and unify your go-to-market efforts.

Articulating Your Promise

How can you tell stories in a compelling and memorable way? Study some of the masters, like Steve Jobs. When he unveiled the iPhone in 2007, Jobs described it as: "An iPod, a phone, and an internet communicator . . . These are not three separate devices. This is one device."[16] This famous keynote, combined with Apple's consistent messaging across PR, marketing, retail, and their website, brilliantly communicated how Apple reinvented the phone. It remains a classic case study in storytelling, worth watching in full.

Consider some of the world's most effective and memorable ad campaigns:

"1000 songs in your pocket." (iPod)
"15 minutes could save you 15%." (Geico)
"Intel inside" (Intel)

The common thread is that the customer promise is a simple, single-minded articulation of why this brand and product are compelling. Campaigns grounded in a customer promise are most effective for new customer acquisition and driving repeat purchases and retention. A study published in the *Harvard Business Review*, based on 2,000 ad campaigns from the World Advertising Research Center (WARC)[17] database from 2018 to 2022, uncovers this insight. Your customer promise drives growth. It connects brand and performance marketing efforts to achieve dual growth, as we'll explore further in chapter 7.

Your messages must adapt to where customers are in their journey and how they'll encounter you. For any innovation, prioritize clarity over cleverness, as customers need to understand your value before anything else resonates. Clearly define what you offer, who it is for, and why it matters. Focus on consumer benefits and use cases, showing how your innovation makes their lives easier, better, and more delightful.

This principle became clear when I experienced Apple's approach firsthand. I'll never forget the days and nights I spent at a warehouse in LA, producing iPhone commercials for global markets. Being at the forefront of many new product launches taught me the rhythm of the creative process and how to build emotional connections with customers through the product experience.

Educating Your Customers

Many tech products solve problems people don't know they have or create entirely new categories. Many innovations don't gain traction because people don't understand how they can benefit their lives. Effective product education is essential to drive awareness, interest, and behavioral change.

For instance, when Apple launched the App Store, the concept of mobile apps was new. The company's "There's an App for That"[18] campaign educated people about the value of apps, showcasing how you could use iPhones to do almost anything you wanted to do, like knowing where to park a car and checking snow conditions on a mountain. Each app was carefully curated to demonstrate the use case and product experience, changing the way people thought about the power of the iPhone and making it easier to grasp. The messaging was simple yet highly effective.

The key was showing, not telling. Advertising simply showcased the product experience in the most ownable and memorable way. For example, we captured the details of a new user's interaction with the device, like pinching to zoom a photo, scrolling a news article, looking at a map—the hand gestures were all part of the communication. All these details added up to tell a powerful story about the new service.

To sustain and continue to drive interest, your campaigns need to stay relevant to customers and culture. When I worked on iPhone 6 in 2015, we produced the first "Shot on iPhone" ad campaign, which would later become the longest-running campaign in Apple's history. The business goal was to sell more iPhones, but instead of talking about the specs of its new, more advanced camera, we demonstrated photos that were being taken by iPhone customers globally. At the time, lots of people were sharing photos on Instagram using #shotoniphone, so we leveraged that organic behavior and turned the world into a gallery. We even put some on billboards.

Product education is about changing people's perspective, and it can be inspiring too. The campaign not only demonstrated the new iPhone's features in the most compelling way but also changed how millions of people thought about photography and their own creative potential. It showed how the iPhone could unlock creativity in everyone, creating cultural and emotional impact. This structured product marketing framework followed Jobs's legacy of treating advertising as an extension of the product. It's a prime example of how brand, product, marketing and advertising efforts worked together to support the launch.

Today, new products and brands must find compelling ways to communicate value, especially when the offering is new or disruptive. Radically innovative services, such as Waymo for autonomous driving or Robinhood for banking services, require deeper education and storytelling to overcome high switching barriers. The journey from awareness to adoption is much more complex than purchasing a new pair of socks online.

When you consistently tell stories about your promises and educate customers through creative narratives across the entire customer journey, you build the momentum that draws customers in and keeps them coming back.

Mastering the Four Stages of a Sequenced Launch

A successful launch doesn't happen all at once; it's built in stages. A phased launch approach helps you roll out GTM activities strategically with clear goals, expectations, and measurement criteria. Business priority, product readiness, target audience, and acquisition economics will all impact your strategic timing and scope to scale up. Here's a four-stage guide to help you win in the market.

Prerequisites for Sequenced Growth

1. **Business Priority**
   Define whether your goal is achieving product-market fit, expanding customer acquisition, or driving profitability. Align marketing scope and acquisition budget with business strategy.

2. **Product Readiness**
   Assess your product experience and its readiness to meet customer expectations. Test and optimize before activating large-scale marketing efforts to avoid disappointing early customers.

3. **Target Audience**
   Clearly define your core audience. Start with a niche market and expand gradually. Mass advertising may be premature before achieving a critical mass of customers.

4. **Acquisition Economics**
   Collaborate with finance to establish acquisition cost

guardrails. Test different channels and refine your strategy to balance investment and ROI.

Four Stages of a Sequenced Launch

STAGE	GOAL	FOCUS AREAS
Pre-Launch	Build your product and brand experience	Brand and product experience, customer insights, launch readiness
Launch	Validate product-market fit	High-intent customers, product-market fit, customer feedback
Post-Launch	Dominate a niche market	Optimize channels, customer acquisition, product education
Scale	Expand to accelerate acquisition & retention	Sustainable customer acquisition and retention, long-term financial viability

Stage 1: Pre-Launch

At this stage, focus on building a brand foundation, refining the product and customer experience, and gathering customer insights to ensure readiness for launch. You'll assess marketability and product readiness and deliver launch requirements such as naming, graphics, landing page, and marketing content, which we have covered in earlier chapters.

Stage 2: Launch

It's vital to validate product-market fit and gather early signals and customer feedback, which can be used to improve the product experience and inform the GTM strategy. At launch, a complex channel plan may not be necessary. When putting together your GTM activities, consider your "must-haves" and "good-to-haves" to make prioritization decisions. Keep in mind that the launch is only the beginning of the journey, and it will take time to grow a successful product.

Stage 3: Post-Launch

After establishing the product-market fit, your marketing efforts can focus on dominating a niche market and building your reputation before expanding to mainstream customers. Continuous marketing is important to gain momentum and drive growth.

Stage 4: Scale

When you scale, the channel plan will become more complex and sophisticated. You need to consider a range of efforts across direct-to-consumer acquisition (organic and paid) and traditional distribution channels to generate sustainable demand and growth to achieve financial viability. In chapter 7, we'll dive deep into your scaling strategy.

───────────────── **Chapter Summary** ─────────────────

Pull customers in rather than push your product onto mass markets. Go-to-market efforts often fail without a strong foundation and proper sequencing. Common pitfalls include misalignment with target audience, lack of marketing, overly broad targeting, and premature scaling.

Your foundation before launch sets the stage for product-market fit. The early foundation you lay in positioning strategy, brand identity, and product experience creates momentum for faster market adoption and lower customer acquisition costs.

Find your product-market fit within a niche. Establish your unique value with a focused audience and win your niche first. Iterate your product experience and go-to-market tactics based on customer feedback.

Build momentum through a sequenced customer acquisition strategy. Win high-intent customers first, optimize owned channels before scaling paid acquisition, and improve acquisition economics to prove the financial model.

Effective storytelling pulls customers in naturally. Storytelling isn't just creative expression—it's a strategic tool to draw customers in. Ground your messaging in clear customer promises, focus on product education to overcome adoption barriers, and consistently demonstrate value.

A sequenced journey enables sustainable growth. Success comes from systematically moving through the pre-launch, launch, post-launch, and scale phases, ensuring a strong foundation that solidifies your market position over time.

Chapter 7

Scale Your Brand to Accelerate Acquisition and Retention

*From Growth at All Costs
to Meaningful Customer Relationships*

Many companies fall into the trap of pursuing growth at all costs, relying on performance advertising to "buy" customers while neglecting customer retention and trust. But sustainable growth is about forming deeper customer connections that fuel both short-term acquisition and long-term loyalty.

This chapter explores key challenges and opportunities in scaling, from integrating brand and performance marketing to deepening customer relationships. We uncover fundamental elements of embedding brand power throughout the customer lifecycle to foster retention, advocacy, and lasting success.

From Trust Crisis to Comeback:
Robinhood's Reinvention for Growth

What happens when millions of users flood your trading platform only to find out that you've taken down the BUY button? Public outrage and an SEC lawsuit.

On January 28, 2021, the financial services startup Robinhood became the focus of a controversy that triggered one of the most significant brand crises in fintech history. Inspired by the WallStreetBets movement, millions of online investors rushed to Robinhood to trade GameStop, AMC, and other widely discussed "meme stocks." In response, Robinhood made a surprise move to remove the BUY button for those stocks. "In light of recent volatility, we are restricting transactions for certain securities to position closing only, including $AMC, $BB, $BBBY, $EXPR, $GME, $KOSS, $NAKD . . ." Robinhood said in a company blog post.[1]

The restrictions caused widespread outrage, as people who hoped to make a quick killing on these meme stocks were furious or devastated. Why did Robinhood take down the BUY button without warning its users? Negative headlines quickly emerged:

"The GameStop Aftermath: The Rise of the Anti-Robinhoods."[2]

"In GameStop Saga, Robinhood Is Cast as the Villain."[3]

Conspiracy theories also floated around. It was a nightmare for Robinhood.

As CEO Vlad Tenev later explained, the company was caught by surprise when NSCC (National Securities Clearing Corporation) demanded $3 billion[4] in cash reserves to mitigate the risk with meme stock trading, and it was a response in the interest of the company and existing users. But without timely customer communication to explain the reason, the GameStop incident eroded public trust, ultimately resulting in a $70 million[5] fine for misleading customers and system outages.

The financial services industry is highly complex and regulated, and it requires more customer trust than other industries. When you manage people's money, you must earn their trust first. The crisis also tested Robinhood's resilience and helped them learn the importance of transparency, customer communication, and brand equity.

How did the company bounce back? How did they reinvent themselves to find growth again? It took Robinhood years to regain customer confidence and reinvent itself to thrive again.

Rapid Growth in the Early Days

Robinhood was launched in 2015 by Vlad Tenev and Baiju Bhatt, who met while studying physics at Stanford University. They had an idea for creating a new trading and financial services platform, recognizing an opportunity that the new generation didn't feel loyalty to old financial companies such as Fidelity and Charles Schwab.

With a mission to democratize finance for all, Robinhood was born. Its initial value proposition was focused on three core benefits: a mobile-first app with an intuitive user experience, commission-free trading, and a user-friendly brand. This modern approach and the commission-free benefit were lucrative to young investors, particularly first-time traders, because at the time, brokers such as E*TRADE and Charles Schwab typically charged $4.95 to $6.95 per trade for stock and ETF transactions. These fees represented a significant cost for active traders.

In the year prior to the launch, Robinhood went to market with an invite-only referral program. It instantly went viral on Reddit, where many interested fans wanted to be the first to take advantage of the commission-free benefit. Nearly a million people joined the waitlist in the first year. In the early days, Robinhood's growth formula was to acquire as many first-time users as possible. By 2019, Robinhood claimed to have more than 10 million users.[6]

Robinhood's growth momentum posed significant competition to traditional financial services. In October 2019, E*TRADE, Charles Schwab, and a few others all dropped their commissions to zero. This was great for customers, but Robinhood lost its competitive advantage. "Commission free" was no longer a differentiator. Fortunately, Robinhood still had a distinctive brand and an intuitive app. While its early growth was impressive, Robinhood's journey *after* the GameStop crisis in 2021 is actually more interesting.

Reinventing to Find Sustainable Growth

In 2022, Robinhood faced another brutal reality—nearly half of their customers left the platform after their 2021 peak. The team had put years of work into driving growth, only to discover that half of them were now gone. Perhaps worse, one would normally expect the most active customers to be the most loyal, but Robinhood had the opposite situation—their more active and sophisticated customers reported the lowest satisfaction.

These huge growth challenges forced the company to fundamentally reconsider its entire business. Robinhood's analysis revealed a critical disconnect: Their products were built around first-time customers, not advanced and active users, but the most resilient revenue came from advanced active users. This insight pushed Robinhood to rebuild its business around sophisticated users rather than trying to appeal to everyone.

The company had to shift its focus from acquisition to engagement and retention—pivoting to improve experiences for its most advanced customers while meeting its own financial needs. As CEO Vlad Tenev explained, Robinhood evolved its strategy, product offerings, and team to address the changing financial needs of its core audience of millennials. Robinhood diversified into eight business lines such as retirement accounts, high-yield cash suite balances, security lending, and others that generate over $100 million in revenue. The new strategy strengthened its financial position for the future.

In 2024, Robinhood's stock rose and tripled from $8 per share to $24. As Andrew Reed, a partner at Sequoia Capital, described Robinhood's journey, "It wasn't the decisions of 2024 that have caused the stock to triple. It was the decisions of 2021, 2022, 2023—the hard work, the cultural resets, the trust-building, the focus on customers and on products."[7]

Why Brand Power Accelerates Acquisition and Retention

As we see from Robinhood's story, a strong, durable business is rarely built through modern technologies alone. Sure, good tech is essential, but sustainable growth relies on differentiation, value delivery, and trust, which form the foundation of your brand power.

- **Differentiation:** This is why people choose you over others in the first place. Your unique difference helps attract customers and delivers that first transaction. In earlier chapters, we discussed extensively the importance of building a differentiated brand and product to attract the right customers who align with your value proposition.
- **Value Delivery:** This is why customers engage with your product and keep using it over time. Once you acquire new customers, you need to continue to fulfill their needs and provide value, over and over.
- **Trust:** This is the glue that connects your products with customers and the catalyst for lasting customer relationships. It's vital for acquiring and retaining customers.

These three elements mirror your growth journey from a customer's perspective. For example, an innovative and differentiated product pulls customers in initially (acquisition), value delivery keeps them engaged (engagement), and trust transforms those transactions into lasting relationships (retention).

Robinhood's journey illustrates how building a distinctive brand early on, coupled with intuitive customer experiences, helped them take off and achieve early success. But in driving rapid growth, they neglected two foundational elements: value delivery and trust. This forced them to reconsider their approach and go back to the core fundamentals of sustainable growth: finding the right customers, building the right products to provide value, forming deeper connections to drive meaningful engagement, and actively prioritizing trust and reputation to enhance loyalty.

When you apply differentiation, value delivery, and trust across your entire customer journey and product life cycle, you gain brand power that naturally accelerates both acquisition and retention. Let's dig deeper into five key ways to do so:

- Driving the right growth, not just any growth
- Acquisition: integrating brand and performance marketing
- Retention: strengthening customer relationships beyond CRM
- Amplifying organic growth through community and advocacy
- Building trust as your ultimate growth multiplier

Driving the Right Growth, Not Just Any Growth

Robinhood's story exemplifies many of the challenges startups face when they scale. Durable growth requires a long-term perspective, yet many companies fixate on short-term results. The startup model emphasizes speed and rapid growth, often at the expense of long-term planning. Startups face pressure to prove their business models and profitability to investors, prioritizing survival over strategic brand building. This creates a tension between immediate sales and sustainable demand and brand building.

Even mature companies face similar pressures, with public companies needing to meet quarterly earnings to satisfy shareholders, leading to a strong focus on immediate financial results. During times of economic uncertainty their risk aversion increases, and marketing budgets are often the first to be cut. Even with limited marketing budgets, the focus usually is on performance marketing to drive immediate results.

The opportunity lies in leveraging every acquisition and retention tactic to enhance the brand's equity and strengthen customer relationships, setting a solid foundation for long-term success.

If you are currently thinking about growth, you must avoid a common mistake many companies make to focus on hyper-growth: neglecting engagement and retention. The truth is, driving the right growth is more important, and it means more than just new customer acquisition. Customer engagement and retention are equally critical. Acquiring new customers only to lose half of them is not sustainable, as we saw from Robinhood's growth journey.

In scaling a business effectively, you need to look at growth holistically across acquisition, engagement, and retention to assess the health of the business.

"Let's run some ads" is not a strategy. You need to understand what's driving growth and focus on the key inputs to drive the outputs you want.

Who are your most valuable customers? Who are your less valuable customers right now who might still be meaningful in the long term? Who is most likely to churn out? These are all important questions you have to ask if your goal is profitable growth.

Harvard Business Review highlights an overlooked key to successful scaling strategy: focusing on profitable growth rather than just any kind of growth. This emphasizes a critical middle stage called *extrapolation*, which sits between the *exploration* and *exploitation* phases: "During this stage start-ups pursue two goals. The first is to confirm the extent to which product-market fit shows that there is demand for the company's offering. The second is to achieve what we call *profit-market fit*—to demonstrate not only that the venture can ramp up revenue rapidly but that every new customer brings in additional revenue and incurs only marginal cost—the key to profitable growth."[8] This requires a proven monetization approach, an effective strong go-to-market strategy, and a few other factors.

Three Stages of Venture Growth

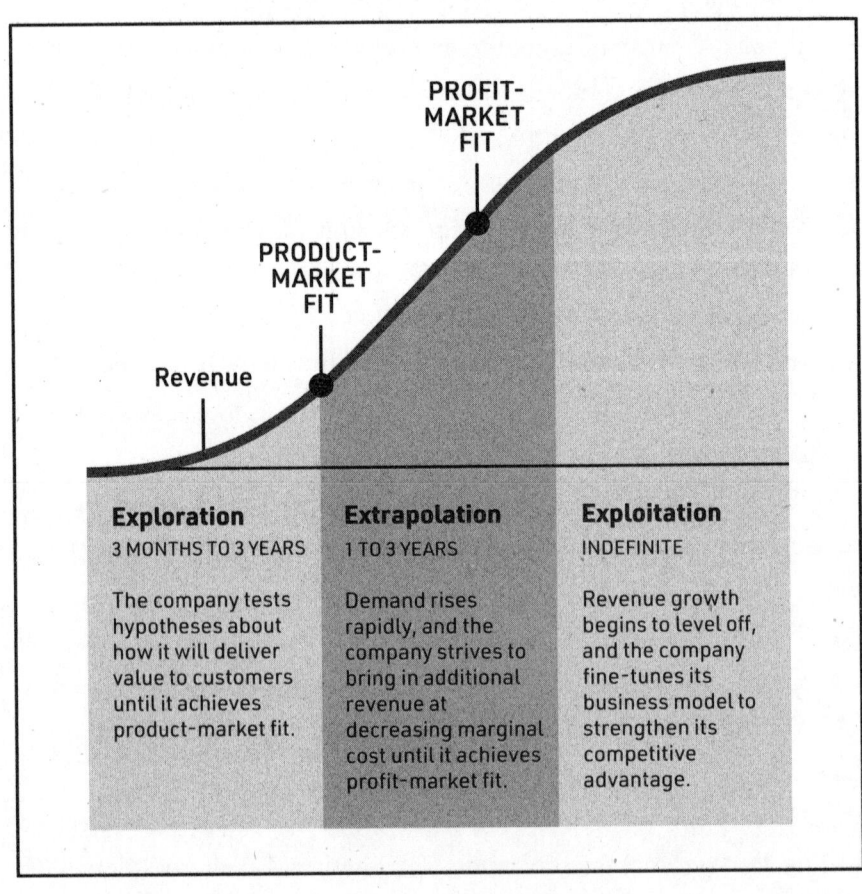

PROFIT-MARKET FIT

PRODUCT-MARKET FIT

Revenue

Exploration	**Extrapolation**	**Exploitation**
3 MONTHS TO 3 YEARS	1 TO 3 YEARS	INDEFINITE
The company tests hypotheses about how it will deliver value to customers until it achieves product-market fit.	Demand rises rapidly, and the company strives to bring in additional revenue at decreasing marginal cost until it achieves profit-market fit.	Revenue growth begins to level off, and the company fine-tunes its business model to strengthen its competitive advantage.

In driving profitable growth, you will often face enormous pressure. I experienced the pain and struggle a few times when working to scale Ring, Key, and Pharmacy and learned the intricacy of trade-offs. I vividly remember countless meetings to debate about where to invest, where to cut costs, and what the unit economics were to improve the health of the business. I didn't love those hard conversations, but they were necessary to figure out how to drive profitability and long-term viability. In those moments, you can see how every initiative you own must make an impact. And you must make the strategy, initiatives, and impact clear. If you can't explain the impact, your initiatives will often get cut.

This is also a moment when marketing plays a crucial role, serving as the voice of customers and providing deeper insights into the metrics you review. Oftentimes, these nuances can help you make better decisions, as you balance trade-offs between short-term acquisition and retention versus long-term brand building.

Acquisition: Integrating Brand and Performance Marketing

Over the last decade, you may have heard about the debate between brand marketing and performance marketing. Companies heavily invested in lower-funnel advertising channels such as paid search to drive direct response, and many teams struggled to manage the integration of brand and performance marketing.

The debate around performance versus brand is problematic because this framing positions them against each other. This comparison treats performance and brand as two opposite and competing initiatives, which is like asking you to choose between your left hand and right hand. Brand building and performance marketing should work together for greater efficiency. Furthermore, this comparison frames brand building in the context of paid advertising, whereas you have many more levers than just advertising to build the brand.

As a study from *Harvard Business Review* puts it, "It's wrong because both performance marketing and brand building impact current revenue and long-term value. It's dangerous because it asks CEOs to accept less of a good thing (say, demand conversion from performance marketing) to make room for more of another good thing (long-term value growth from brand building)."[9] This perspective highlights the urgency of taking an integrated approach and avoiding unhealthy competition for budget allocation.

One of the most compelling examples of this integrated approach comes from Airbnb's dramatic transformation during the COVID-19 pandemic.

How Airbnb's Brand Resilience Fueled Sustainable Growth

In 2020, a global pandemic changed the course of Airbnb forever, just as the company was preparing for its IPO. COVID-19 caused global travel demand to stop, and Airbnb lost 80% of its business in just eight weeks. It was an existential crisis for a travel company that had previously enjoyed steady growth.

"Is this the end of Airbnb?"

"Is this the end of travel?"

These were the kinds of headlines dominating the news that spring. In the face of tremendous doubts, what had been the most promising IPO of the year suddenly seemed on the verge of collapse. With global and long-distance travel demand plummeting, Airbnb restructured its business, adopted a more sustainable cost model, focused on its most essential initiatives, and made the painful decision to let go of nearly 25% of its workforce in 24 countries.[10]

It also quickly pivoted its product offerings to meet what customers were now looking for: safer local travel. While many people didn't want to risk their health for long-distance trips, they were more open to nearby getaways they could drive to, avoiding the exposure of planes and trains. They also preferred staying in Airbnbs rather than hotels to avoid crowds in public spaces. A swift shift to drive customer engagement toward local travel was instrumental in helping turn around the business and avoid the risk of bankruptcy.

Another major decision was to pause all paid marketing and advertising during the worst months of the COVID crisis. When ad campaigns were halted in 2020, Airbnb's marketing team redirected their effort to analyze past marketing investments and gather learnings. To their surprise, Airbnb's website traffic rebounded to 95%[11] of 2019 levels *without* any marketing spend, and by Q4 2020, they even discovered that nearly 90% of Airbnb's website traffic was coming from people who visited the site directly or through unpaid channels, rather than by clicking on ads. This insight confirmed what many brands struggle to measure and quantify: the power of brand as the primary driver of sustained demand and growth.

For years, like many digital-first companies, Airbnb had followed the trend of investing heavily in performance marketing to drive paid traffic. In 2019, the company spent $1.14 billion on marketing,[12] with the majority allocated to performance advertising channels such as paid search. However, internal analysis revealed that the investment on performance marketing channels hadn't driven as many results as they assumed.

With this insight, Airbnb made a strategic shift post-COVID, scaling back investment in lower-funnel performance marketing and doubling down on brand-building efforts. In 2021, the company resumed paid marketing but reduced its overall marketing spend from 34% of revenue (pre-COVID) to just 20%.[13] By 2022, Airbnb generated $8.4 billion in revenue[14] and grew 40% year-over-year, which made 2022 its first profitable year.

Airbnb achieved this remarkable turnaround—recovering after losing 80% of the business, reducing marketing costs, improving efficiency of its marketing investments, and driving future demand and growth—all within a highly competitive travel category while other services outspent Airbnb significantly. For example, in 2022, Expedia Group and Booking Holdings spent $6.1 billion and $6 billion,[15] respectively, on sales and marketing expenses. Yet, as Airbnb's chief financial officer Dave Stephenson described it, the company's shift in marketing strategy and investment from 2020 to 2022 was "incredibly effective."[16] Airbnb's CEO, Brian Chesky, explained it during his November 2022 earnings call: "We think of performance marketing as more of a way to laser in to balance supply and demand rather than a way to just purchase a large number of customers."[17]

How was this possible? Airbnb possessed an important asset that's hard for competitors to replicate: a strong brand. This pivot would not have been feasible if Airbnb hadn't spent years building its brand and strengthening its product offerings. Its brand resilience wasn't accidental, but it was the result of deliberate, long-term brand-building efforts throughout the years. This included a pivotal repositioning and rebranding initiative in 2014 to center its brand and product offerings around the core idea of *belonging*, which we discussed in chapter 4. These efforts laid the foundation of Airbnb experiences and deepened its connections with customers.

Many teams struggle to integrate brand and performance marketing, often inadvertently positioning them against each other. However, this approach can hinder new demand creation and sustainable growth. Airbnb recognized the importance of building a strong brand and product offering to attract the right customers and moved the company away from simply buying traffic.

Airbnb's shift from performance marketing to brand marketing reflects a deeper understanding that sustainable growth comes from building your brand differentiation and forming meaningful connections with customers, not just driving traffic and transactions. This integrated approach ensures that every dollar spent not only delivers short-term results but builds foundations for the long-term. Every message, every acquisition moment, every transaction can strengthen your meaningful differentiation, deliver value, and earn customer trust.

An Integrated, Full-Funnel Campaign Approach

How do you create an integrated approach where both brand building and performance marketing work together? The key lies in implementing a full-funnel strategy where your investment mix depends on your business priorities and stage to balance long-term and short-term growth goals.

Performance marketing typically focuses on immediate response metrics like click-through rates, conversion, and cost per acquisition. Brand marketing tracks these metrics but also looks at awareness, perceptions, and loyalty. These efforts can and should work together. Your goals and investment might vary, but you can create this integration by focusing on two key elements: communicating your differentiation (what makes you uniquely valuable to customers) and ensuring consistent creative execution.

First, the foundation of building a brand is about establishing a unique positioning in the minds of potential customers, and this is how your differentiation becomes relevant and meaningful to customers. Your messaging should be grounded in a customer promise, even in performance channels such as paid search and digital display. You can identify and validate your most effective messages through a Structural Equation Model study, which uses a set of statistical techniques to map out the complex relationships between your messaging elements and business outcomes. For example, it gives you quantitative insights into how different messaging variables contribute to acquisition and retention, such as whether emphasizing price or delivery speed has a stronger impact on acquisition versus retention. Your customer promise helps drive sales and strengthen perceptions simultaneously.

Second, the selling points can also become brand-building moments. When you apply consistent targeting, messaging, and creative work to different channel environments, you create a unified experience for customers to discover and engage with you. For example, consider how Apple maintains consistent brand expression whether you're visiting their website, visiting an Apple Store, or browsing the Apple section at Best Buy. The brand is instantly recognizable across all touchpoints, turning each selling opportunity into a moment to foster a great customer experience. We can't underestimate the power of consistency to reinforce your brand as you scale.

So, this full-funnel, integrated approach creates a direct link between brand equity growth (your brand differentiation) and customer acquisition outcomes by thoughtfully managing two key elements. One is channel activation—creating an effective channel mix to target the right customers at the right moments with the right cost. The other is creative execution—developing messaging and creative work that's tailored for channel activation while consistently reinforcing your brand. When these elements work together, every channel activation improves perceptions while driving acquisition.

This integrated, full-funnel approach also applies to how you measure the effectiveness through both short-term and long-term metrics together, which we'll explore in the next chapter.

- Short-term: Click-through rates, conversion rates, cost per acquisition
- Mid to long-term: Awareness, perceptions, consideration

Integrating Performance and Brand Marketing

CHANNEL ACTIVATION

Targeting the right customers
at the right moment with the right cost

$+$

CREATIVE EXECUTION

Developing the right messaging
and creative tailored for channel activation

$=$

CUSTOMER ACQUISITION

Achieving the double duty of selling
and building brand equity

For example, when we marketed Amazon Pharmacy in lower-funnel channels across paid search, display, and social, we targeted high-intent customers and used messages such as "Prime members get 80% off" and "Your medications delivered to your door." These messages around price and convenience not only drove consideration and purchase but also improved perceptions that Amazon Pharmacy is convenient and offers low prices. This approach achieved the dual goals we discussed earlier.

Retention: Strengthening Customer Relationships Beyond CRM

Many of you may use customer relationship management (CRM) as a sales tool. "Let's send another email to ship more units!" That sounds familiar, right? We often forget that emails, messages, and app notifications are among the most personal communication channels we have with customers.

Retention isn't just about keeping customers and driving repeat purchases. It's about continuing to fulfill their needs and delivering value. You need to shift the thinking from "How can I sell you more stuff?" to "How can I add more value for you?" This value—both emotional and functional benefits—deepens relationships and creates long-term growth.

Every customer touchpoint can become an opportunity to reinforce your value: An order confirmation transforms into a memorable interaction, a notification evolves into a helpful tip, and customer data becomes the foundation for personal experiences.

Turning Transactions into Memorable Moments

Today, people's inboxes are filled with junk mail and upselling messages; it's no wonder that many take any opportunity to opt out of emails. But email can still be a powerful channel to reinforce your brand and connect with customers more deeply. How do you make your customers appreciate your emails instead of dreading them?

Flamingo Estate, an online lifestyle retailer that sells farm-grown products and experiences, is a good example of bringing its brand to life and crafting personalized and delightful moments in digital environments. Founded by Richard Christiansen in Los Angeles, Flamingo Estate positions itself to be about radical pleasure, selling products from extra-virgin olive oil to persimmon vinegar to sage candles directly from the garden. During the holiday season, I ordered a seasonal gift box for a friend. From the transaction email to the notifications about the delivery, every communication was well-crafted. In fact, I could feel the simple pleasure from the garden:

"The Garden's growing towards you. Pleasure is on its way."

"Hello, Pleasure. Let's get to know each other."

Their email communications delight people's senses through evocative language and storytelling that embodies their "radical pleasure" brand positioning. Even in simple order status notifications, the storytelling makes you feel good about your purchase. This feeling shines through in their thoughtful communications. Your email's tone and style can reinforce your brand's personality. When you inject feelings into your email, even mundane communications like transaction and delivery notifications can become memorable.

Designing Customer Communications

A common challenge for scaling is the lack of coordination to manage the end-to-end customer communication across cross-functional teams. In many organizations, the product team manages a part of the customer journey (often notifications), and the marketing team manages another part of the journey (acquisition and retention). These teams often don't coordinate well to align on their communications tone and style.

For instance, I remember a conversation I had with a product leader about notifications. His team received an urgent request from senior leadership: "Clean up the notifications!" What was the problem? No one was overseeing the entire customer experience, and the product team was sending out an overwhelming, annoying number of customer notifications.

This product leader assigned a dedicated manager to clean up all the customer-facing notifications. They also redesigned the entire process for customer communication, which led to improved adoption rates and customer satisfaction. This example serves as a reminder that notifications, while seemingly minor, can add up to impact customer experiences.

Companies may put extensive effort to support new product launches and more prominent customer experiences, but ongoing notifications are often neglected until they create problems down the road.

Today, any product can generate a lot of real-time notifications, via an app, text messages, and email. If not done properly, you risk frustrating customers and making them want to opt out of the communication. For example, imagine launching an AI-powered personal trainer app. You need to acquire new customers to drive app adoption and growth, and you also want to onboard and engage existing customers to remind them they need to exercise at their preferred time. If customers already churn out, you might send out messages to win them back. Instead of sending the same workout reminders to new customers and existing members, it's essential to map out the entire customer journey, understand where they are, and design customized communication at that moment to deliver value and drive meaningful engagement.

The key is auditing your entire communication flow: What messages are customers receiving? How often? Are they adding value? This may sound trivial, but small things add up—and small details differentiate good products from great brands. It's about attention to detail and *consistent* care for customers.

Deepening Customer Connections Through Data

As you collect data from various channels such as email, social media, website, and app, you can learn to anticipate customer needs and behaviors better. Sometimes, data can deliver insights that help you create even more personalized experiences.

For example, Spotify excels in designing highly personalized, relevant experiences that excite and surprise their customers. Its famous Wrapped campaign[18] every December demonstrates how a brand can leverage data collection to create personalized content to engage existing customers, acquire new customers, and build brand love.

Beginning in 2015,[19] Spotify transformed listening habits and product usage data from more than 500 million users into an integrated year-in-review marketing campaign that celebrates Spotify's personalized experience. The campaign features real Spotify customers and highlights their unique listening habits.

The Wrapped campaign was brought to life across all touchpoints, including Spotify's iOS and Android apps, website, email, social media, and paid advertising channels such as roadside billboards. It was highly successful and viral, as Spotify made it easy for users to share on social media and encouraged user-generated content to drive social media engagement. It also fostered a sense of community, making people who don't have Spotify feel left out. The combined power of social media, user-generated content, and community spirit deepen customer relationships, amplify the brand, and make it culturally impactful.

Amplifying Organic Growth Through Community and Advocacy

You might be thinking, *I don't have a big marketing budget to run campaigns, so how does this work for me?* A meaningful brand isn't built through marketing spend alone. Even substantial marketing investments won't create lasting success without consistently providing differentiation, fulfilling customer needs, and earning trust. Many companies don't heavily invest in paid media but rely on PR, community building, word of mouth, and other organic strategies to generate exposure beyond what traditional advertising could achieve. This can become an effective flywheel to drive stronger customer engagement and loyalty, as we saw with Spotify Wrapped.

Public Relations

PR is powerful when it authentically showcases what makes your product and brand unique. It helps raise awareness, reinforce trust, and earn media coverage.

When Airbnb reduced its investment in performance marketing to focus on brand building, PR became one of the most important marketing channels to keep Airbnb top of mind for travelers. The company's focus on pop culture and experiential marketing helped drive positive press coverage and reach new segments and audiences. For example, Airbnb partnered with Warner Bros. to promote a listing of the Barbie Dreamhouse on Airbnb during the summer of 2023, to leverage the release of the eagerly anticipated *Barbie* movie. This experiential marketing was newsworthy, generating over 13,000 media hits and 250 million[20] impressions on social media—more than twice as many impressions as Airbnb's IPO announcement.

When Airbnb launched a new product offering called *icons*, which features entertainment and cultural experiences such as Pixar's *Up* house, X-Mansion from X-Men, Musee d'Orsay in Paris, and the Ferrari Museum in Italy, PR continued to be an effective channel to drive excitement about these new experiences, which people could book through Airbnb. Even though these icons offerings didn't drive significant revenue, they helped Airbnb generate massive news coverage and significant exposure beyond what traditional advertising could achieve with the same level of investment.

For startups without big marketing budgets, it takes creativity and unconventional thinking to generate word of mouth and earn organic media coverage. In the early days, Liquid Death took an irreverent, humorous approach to branding that created memorable moments and inspired organic media coverage. The company didn't even have a product yet when they posted their first video on Facebook, which generated three million views before any water was available for purchase. Similarly, Robinhood's commission-free proposition became newsworthy before they even launched their service. Their invite-only referral program went viral on Reddit and attracted nearly one million people to join the waitlist in the first year. The key for both was having something genuinely newsworthy and powerful that gave the media a reason to spread the word.

Community Building

Community building creates stronger customer connections and drives word of mouth. Tapping into superfans can be rewarding as they become your biggest advocates and help promote your products and services.

For example, remember how Notion attributed its early success to its superfans and community efforts? Notion users loved the product, so they talked about it in other online communities. The company identified natural behaviors and developed strategies to nurture and support the customer passions that they observed. When early users created and shared user-generated templates, for instance, Notion provided support by creating a template gallery on their community page. This fostered a sense of community contribution while showcasing the product's versatility. Notion encouraged user creativity and created a network effect—the platform became more valuable as more users contributed.

Notion also started an ambassador program to bring together superfans from around the world. These ambassadors were volunteers who wanted to teach and share Notion with others. They organized local events and meetups, led local and virtual communities, and helped create educational content to share best practices. The Head of Community at

Notion wrote on the program's page: "We have so many passionate makers, builders, creators, all over the world who are inspiring others with their approach to Notion, and we're excited about bringing these people together to share their stories. Ambassadors are so core to who we are at Notion, so we wanted to design a program that celebrates and empowers them to accomplish even more."[21]

Notion's focus on community directly contributed to its success. As of 2021, 90%[22] of its growth was organic, and community building helped the company scale its product to more than 100 million users.

Product as Marketing

Sometimes, your product itself can become a powerful marketing tool. That's the case for Ring. The videos captured through Ring cameras represent the actual product experience in action. When customers share footage of package deliveries, wildlife encounters, or neighborhood activities through the Ring Neighbors app, they're demonstrating the product value better than any traditional advertising could. Ring transformed this user generated content (UGC) into an organic content marketing machine that drives word of mouth and builds community trust. It's an effective way to convince people to buy your products.

Ring has cultivated a community of passionate users who want to share video content captured through their doorbells and cameras with other neighbors, to make them feel connected to their neighborhood. From seeing what's happening at the front door to capturing bears and coyotes in the backyard, to documenting the kindness of strangers to help their neighbors, this user-generated content created significant word of mouth and improved awareness of Ring.

Some of the most compelling UGC content from Ring's Neighbors app was featured and amplified on Ring's social channels, PR, and paid advertising. These organic levers worked together with Ring's paid media efforts to maximize awareness and fuel its growth and expansion.

Building Trust as Your Ultimate Growth Multiplier

"Can I trust you?" is one of the first questions people ask when considering a new service. Without trust, transformative services like Airbnb or autonomous driving wouldn't be possible. Healthcare, financial services, and AI products demand significant trust, which needs to be addressed early on. Trust is about helping people overcome the unfamiliar—especially crucial for innovative products that ask customers to change behaviors. Trust also amplifies word-of-mouth referrals, builds strong communities, and drives customer advocacy.

Taking the Trust Leap

When you push boundaries to bring transformative products and services to market, you are asking people to change their habits, whether that means buying prescription medications online, stepping into a self-driving car, or allowing a robot to perform surgery on them. This is often called a "trust leap"[23]—and can be a huge obstacle for innovations. Most people are simply reluctant to try the unfamiliar. As Rachel Botsman, author of *Who Can You Trust?*, explains, "Trust is the confident relationship with the unknown." This definition highlights the fundamental challenge many digital products and services face.

Without trust, we can't imagine a world where we use Airbnb to stay at strangers' homes, or use Uber to get in a stranger's car. Technology has transformed these and many other relationships, creating a paradigm shift in how we connect with products and services. As the future becomes more AI-driven, from self-driving cars like Waymo to robotic surgical systems like Intuitive da Vinci, brand trust becomes even more essential.

For example, Waymo has a big challenge ahead. To achieve mass adoption and scale, all autonomous driving companies need to help customers make a trust leap. Waymo's mission to become the world's most trusted self-driving car will require patience and consistent effort. The company is asking people to make a big leap that requires both rational risk-taking and emotional comfort before they're willing to jump in.

Trust Is a Multiplier for Organic Growth

Trust is a powerful way to build reputation and unlock scalable growth. When customers trust a brand, they recommend it, allowing you to find your growth organically without additional paid advertising spend.

So, what builds trust in innovative products and services? The short answer is *everything*: product design, customer experiences, marketing campaigns, and public policies. It's every small action that helps educate customers, maintain consistent messaging, and demonstrate transparency. It's every action that reinforces reliability and consistency in value delivery.

Building trust often requires giving customers control and helping them see immediate value. When people can understand how something works and feel in control of their experience, they're more likely to overcome initial skepticism and embrace new solutions.

As in human relationships, trust is the foundation for long-lasting connections. Achieving trust was the payoff for Ring's leveraging of user-generated content, Notion's nurturing of a template-sharing community, and Robinhood's dedication to its most important customers. In the end, trust is the multiplier behind organic and continuous growth.

———————————— **Chapter Summary** ————————————

Brand power accelerates acquisition and retention. Sustainable growth relies on differentiation, value delivery, and trust. Your brand power comes from how well you deliver on these elements as they form the foundation that accelerates both acquisition and retention.

Focus on driving the right growth, not just any growth. The opportunity lies in leveraging every acquisition and retention tactic to enhance the brand's equity and strengthen customer relationships, setting a solid foundation for long-term success.

Integrate brand marketing and performance marketing together as a full-funnel campaign. This integrated approach ensures that every customer touchpoint contributes to both immediate sales and long-term brand equity, strengthening your brand differentiation and deepening customer connections.

Strengthen customer relationships. Retention isn't just about keeping customers and driving repeat purchases. It's about continuing to fulfill their needs and providing value over time. You need to shift from "How can I sell you more stuff?" to "How can I add more value for you?"

Amplify organic growth through community and advocacy. Organic strategies such as PR, community building, and user-generated content can become an effective flywheel to generate word-of-mouth exposure, driving stronger customer engagement and loyalty.

Build trust as your ultimate growth multiplier. Trust is about helping people overcome the unfamiliar and take a trust leap to adopt innovative products and services. It also amplifies word-of-mouth referrals, builds strong communities, and drives customer advocacy.

Chapter 8

Iterate After Measuring Impact

From What's Easy to Measure to What Truly Matters

Brand power accelerates customer acquisition and retention, but how can you measure success and make results tangible? In a data-driven environment, it's easy to fall into the trap of what's easy to measure and miss what truly matters to your customers and your business—the deeper indicators that allow you to build for the long term. This chapter explores how you can leverage integrated frameworks and strategies to connect brand power with financial metrics so you can iterate and optimize your business.

Nike's $28B Lesson: When Data Revealed What Truly Matters

On June 28, 2024, right after Nike announced its quarterly financial results, the company's share price plummeted by 20%,[1] making it one of the worst days in its history and erasing $28 billion from its market cap. Nike Direct revenue was $5.1 billion,[2] down 8% on a reported basis due to declines in Nike Brand Digital of 10% and Nike-owned stores of 2%. Massimo Giunco, former senior brand director at Nike, described the crisis on a LinkedIn post, "Nike: An Epic Saga of Value Destruction."[3]

What happened to Nike? Why did a data-driven transformation agenda turn out to be a disaster? The answers provide a cautionary case study.

When Data Told a Different Story

In early 2020, Nike announced a digital transformation strategy called *Consumer Direct Acceleration*. The plan was a reinvention of Nike's business model, powered by technology, data, and direct customer relationships, with the assumption that their DTC channels (Nike.com, the Nike app, and Nike stores) along with digital performance marketing and advertising were going to capture sales that were previously contributed by wholesale retailers. Nike made three major changes:

- Eliminating sports product categories and shifting to "a new, simpler consumer construct of Men's, Women's and Kids"[4] category structure
- Focusing on DTC sales and ending traditional wholesale retail partnerships
- Shifting its marketing model to be data-driven and digitally led

Overall, Nike pulled back from its previous "brand-led" approach to a "sales activation" strategy, with increased investment in programmatic advertising and performance marketing to drive traffic to Nike.com.

Initially, the strategy seemed successful due to the pandemic-driven shift to online shopping and the residual effects of brand investments Nike had made in previous years. But after two years, the formula of using performance marketing to drive traffic and leveraging Nike's direct channels no longer converted traffic into sales. Nike's sales were dropping quarter by quarter. Worse, the constant promotional activities and discounts eroded gross margin, brand equity, product equity, and market share.

After seeing disappointing results, Nike reversed its strategy and increased its presence in third-party wholesale retailers and B2B partnerships.[5] It ramped up brand advertising investment to reinforce its relevance and maintain top-of-mind awareness. Additionally, Nike went back to reorganize its product offerings around core sports categories like basketball, tennis, football, and so on, to realign with customer interests.

A Universal Challenge in Tech

Nike isn't a technology company, but its financial hit reflected a broader trend facing tech products and services: an obsession with short-term, performance-driven strategies and metrics at the expense of long-term brand building. In the rush to embrace data-driven decision-making, many companies prioritize easily trackable metrics, often losing sight of what makes their businesses special and what truly drives sustainable demand.

The short-term mindset is further amplified by the democratization of business infrastructure. Platforms like Shopify empower startups with easy-to-build storefronts, scalable back-end solutions, and data-driven customer management tools. As a result, businesses now have unprecedented control over the entire customer journey, including post-purchase interactions. While AI technologies and digital capabilities have lowered barriers to entry, they haven't eliminated the need for product innovation and brand strength, which are much harder to measure.

Many digital-native brands like Glossier, Warby Parker, Casper, and Dollar Shave Club pioneered direct-to-consumer models, leveraging digital channels and performance marketing to grow quickly. But like Nike, they learned that being DTC and digitally led isn't enough. Sustainable growth requires more than efficient distribution and digital advertising. It requires constant product innovation, strong brand equity, and deep customer relationships.

Nike has always been more than a footwear and clothing company—it's a brand that celebrates sports and everyday athletes. Its core belief that "If you have a body, you are an athlete" has made Nike an inspiring symbol customers love.

Building a brand that resonates with customers is not easy. Product innovation that exceeds customer expectations requires significant effort. Forming deeper relationships with customers and making them happy demands dedication. But it's these hard things that separate good companies from great ones. Even a company with strong brand equity like Nike can fall into the trap of prioritizing short-term metrics, so it can happen to any business.

As sports retail expert Matt Powell put it, "What happened on June 28th on Wall Street is just the result of what was decided four years ago . . . What we saw is an epic saga of value destruction, harming Nike's brand mental and physical availability, in just three years."[6]

This pattern reflects a broader industry challenge: Most professionals only experience one stage of a product's life cycle and few stay long enough to see it through different business stages. People change jobs frequently, move to different companies, and marketing agencies rotate accounts. This fragmentation means few people see how early brand decisions compound over years, creating pressure to focus on the immediate deliverables and metrics.

Nike's value destruction showed the danger of transforming an established business model too radically and too quickly. Efficiency alone is not a substitute for strong brand equity. When Nike moved away from its brand differentiation and strong retail partnerships to focus on direct-to-consumer channels and performance marketing, its business suffered. Nike discovered that customer loyalty wasn't as strong as they had assumed. When it pulled traditional retail partnerships and customers were given more choices, those partners and customers quickly explored alternatives. Brand strength isn't static. To sustain growth, a brand must stay distinctive, relevant, and consistently accessible.

Why Brand Equity Matters and How to Measure It

What is brand equity? Simply put, brand equity captures and reflects customer perceptions, what you're known for, and where you are in the market. As David A. Aaker defines it, brand equity is "a set of assets (and liabilities) linked to a brand's name and symbol that adds to (or subtracts from) the value provided by a product or service to a firm and/or that firm's customers."[7] Remember the definition of brand and how it exists in the mind of customers? Brand equity is the measurable impact of those perceptions on your business.

Having worked with Nike in the beginning of my career, I witnessed firsthand how the company was built from its passion for sports and deep connection with athletes. What made Nike special and unique was its brand, product innovation, and sports communities. That combination became the foundation of Nike's long-term competitive advantage.

Brand Equity Is the Engine for Sustainable Growth

Building brand equity means investing in your meaningful differences. Without it, you become generic, making it hard to stand out in the market. When Nike eliminated sports product categories and shifted to Men's, Women's, and Kids, the company stripped away what made the brand distinctive, the very reason people buy Nike. Its shift from passion-driven sports communities to generic classifications around gender and demographic inadvertently diluted its equity and reduced itself to another footwear and clothing company. This is precisely why brand equity isn't merely a marketing metric but a critical business asset with tangible financial value.

What contributes to brand equity growth? It always starts with product innovation and differentiation to create something customers genuinely value. Today, product experience is the most powerful driver of brand equity because customers form their strongest perceptions through direct interaction with your product. This is why product experience forms the foundation of the *Brand Power Built In* approach. This approach requires a mental shift in how you measure success. Instead of measuring brand and product success separately, this approach connects the dots and shows how they work together to drive business results.

How to Measure It

As we can see from Nike's story, brand equity metrics serve as key indicators of your business and long-term growth. They signal your brand's meaningful difference among potential customers and existing customers, help you understand your position in the market, and provide insights into key barriers and motivators behind your customer acquisition and retention efforts.

- **Awareness:** Do your potential customers know about your products?
- **Perceptions:** What do your potential customers think of your differentiation?
- **Satisfaction:** Are your existing customers happy with your products?

These in turn drive your business performance metrics:

- **Acquisition:** How many new customers do you acquire?
- **Engagement:** How many existing customers actively engage with your products?
- **Retention:** How many existing customers do you retain?

Here's how these metrics are connected and directly impact customer acquisition and retention and business performance. For customer acquisition, potential customers first must know about you before they can buy your products or services; that's why awareness is the first step. They also need to know your unique value proposition and perceive your differentiation in ways you intend. Awareness and perceptions are direct indicators of *why* customers will want to choose your products over the alternative. For customer retention, existing customers need to be happy with your products or services before they recommend you to their family and friends. The key metric is Net Promoter Score (NPS), an indicator of customer satisfaction and loyalty, measured through one simple question: "How likely are you to recommend our product or service to others?"

Brand health metrics are usually collected through surveys. While product usage data and business performance metrics provide immediate feedback, brand health metrics provide a macro view of the business and give insights into the why behind your acquisition and retention data, which we'll explore more in the later part.

There's a common misperception that brand is primarily about creative work or paid mass advertising—the narrow view we discussed in chapter 1. This misperception creates measurement challenges that must be overcome via driving clarity and alignment across your organization. Being specific is particularly important. For example, if you are evaluating the impact of paid advertising, then say that directly rather than using the broader term "brand." This specificity becomes especially critical when measuring integrated marketing campaigns, which people often describe as "brand campaigns," perpetuating these misperceptions.

Measuring Integrated, Full-Funnel Campaigns

Brand equity grows through many initiatives—product improvements, customer service, partnerships, and more. But how do you measure the specific impact of your marketing campaigns? As we've learned from Nike's $28 billion value decline, efficiency alone is not a substitute for strong brand equity, but measuring how brand equity growth translates to business results presents unique challenges. If you lead marketing and invest in mid- to upper-funnel marketing campaigns, you may wonder, "How can I justify my budget and make impact visible?" and "How can I help my cross-functional partners understand this?" As we explored in chapter 7, integrating brand and performance marketing into a cohesive full-funnel approach creates a direct link between brand equity growth and customer acquisition outcomes—but measuring this integration effectively requires a thoughtful framework.

Finding the Right Metrics for the Full-Funnel

How can you measure lower-funnel and mid- to upper-funnel together? Campaign measurement is complex, and finding the right metrics to track can be difficult. You need to align on your primary and secondary goals and metrics for your campaign. If your primary goal is about improving awareness, then the measurement metrics should first prioritize awareness metrics over direct sales results. Vice versa, if the goal is about driving sales, then the measurement metrics should be centered around sales. You need to align on priorities and track multiple metrics across the full-funnel to understand the complete impact.

Campaign Measurement

- **Upper-funnel:** awareness, perceptions, consideration
- **Lower-funnel:** traffic, click-through, conversion rate, engagement
- **Cost and efficiency:** cost per impression, cost per click, cost per acquisition

When you put brand equity, campaign measurement, and business performance metrics together, your measurement gets complex fast. Whether you're in product or marketing, this is where you need to spend time and effort aligning on goals, success metrics, and connections between inputs and outputs from cross-functional teams.

Finding the right metrics can be liberating as it has the power to align the entire organization and create clarity. For example, Airbnb used traffic sources to assess the impact of its brand strength in generating organic demand versus paid advertising demand. Roughly 90% of Airbnb's traffic came direct, without paid ads. Measuring and comparing paid traffic versus organic traffic helped Airbnb measure the effectiveness of their investment. This metric led Airbnb to make the strategic shift to focus on brand building.

Identifying sources of awareness can be a powerful tool for analysis. When I was working on Ring, we used a simple question—"How did you hear about us?"—to measure and track awareness sources. This one-question survey, included on the website, embedded in the product UX flow, or sent via email, helped uncover which awareness channels drove sales. This data enabled the team to attribute sales to specific marketing investments and optimize budget allocation to achieve the highest ROI.

The Attribution Problem:
Siloed Metrics Undermine Brand Success

One of the biggest challenges in measuring full-funnel campaigns is attribution. If you ask CEOs and executives whether customer perceptions matter, most would say yes, of course. But when they face cost overruns and need to trim their budgets, brand-related initiatives are usually among the first to be cut.

This disconnect exists because, while lower-funnel marketing activities can be directly linked to short-term sales metrics, awareness and perception metrics are often not reported instantly, because they require separate surveys. The diffused impacts are harder to attribute to specific business KPIs. This measurement challenge frequently results in brand efforts being undervalued.

The head of brand of one of the Big Five tech companies recently told me that the success metrics for her workstream were about brand sentiments, measured in isolation from business, product, and growth metrics. She expressed serious concerns that this siloed measurement set her and her team up for failure, because her success metrics were not directly linked to broader business metrics. She worried that senior leadership couldn't see the impact of her team's work. My friend's experience is a common industry challenge. Brand measurement is complex because each company uses different metrics, leading to attribution challenges and a lack of standardized, integrated measurements.

Here are some common attribution challenges that hide a brand's true value.

- **Full-funnel complexity:** No single tool can capture the full-funnel data. Lower-funnel metrics like click-through rates and conversion rates don't capture customers' motivations, while each digital ad platform (Google Ads, Facebook, Instagram, etc.) uses its own metrics and dashboards, further fragmenting measurement.

- **Time frame differences:** Unlike direct response metrics, changes in awareness and perceptions take longer to manifest. While click-through rates provide instant feedback, meaningful shifts in customer perceptions develop gradually. This often requires a stand-alone study to link upper-funnel investments to business growth.

- **Business prioritization:** Marketing priorities change and evolve with your business stages. Growth companies might focus on improving awareness and perceptions among specific customer segments, rather than the general population, while mature companies prioritize retention metrics. Since you can't improve all metrics simultaneously across all audiences, you often have to prioritize and make trade-offs to align with your business goals.

Full-funnel metrics, varying time frames, and shifting business prioritization make measurement genuinely complex. If you're feeling overwhelmed trying to piece it all together, you're not alone. Creating a sophisticated attribution model typically requires an experienced data science and analytics team. This work requires time, investment, and team collaboration, and it's not something you can knock out easily.

Despite these challenges, you can't afford to keep measuring everything in silos and you need to look at metrics through an integrated framework. On paper, it's much easier to see that whenever you spend $X on Google paid search or other channels, you generate $Y in incremental revenue. But building strong brand equity isn't nearly as straightforward. It requires a holistic view that considers numerous key variables and applies thoughtful analysis to them.

Connecting the Dots Through the Input-Output Framework

The renowned management thinker Peter Drucker is often credited with saying, "If you can't measure it, you can't manage it." That insight is especially relevant to assess the impact of integrated campaigns. The input-output framework provides a structured approach to connect your activities and business outcomes. You first need to know what to measure and then find a way to connect the dots.

What's the Input-Output Framework?

The input-output framework emphasizes the actions and factors you can directly influence (inputs) rather than focusing solely on lagging financial indicators (outputs). It bridges the gap between brand equity metrics and business financial results by connecting controllable inputs with measurable outputs.

Here's how it works for marketing. Customer acquisition, retention, units, and revenue are output metrics. Things that feed into growth, such as a better value proposition, refined targeting, compelling creative work, and enhanced product and customer experience, directly improve awareness, perceptions, and demand. You can see how your controllable inputs drive outputs of what you track in acquisition, retention, and unit metrics and how they're connected. You should focus your energy on inputs, what you can do across product, marketing, customer support, and other departments to grow the business together.

Of course, business performance results are always the most important. But the challenge is that you may not always connect the dots. For example, if your product isn't adopted by customers as fast as you expected, you need to figure out why—and lower-funnel metrics can't always tell you the full story. Conversely, if your business grows faster than you imagined, you need to understand the key growth drivers and uncover the correlation. In a growth-driven world, it's easy to get

Marketing Input vs. Output Metrics

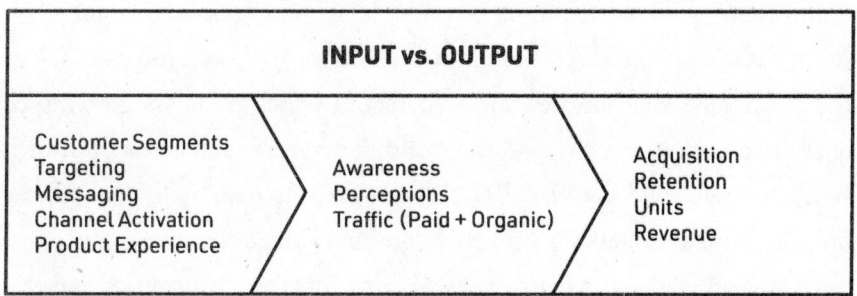

INPUT vs. OUTPUT		
Customer Segments Targeting Messaging Channel Activation Product Experience	Awareness Perceptions Traffic (Paid + Organic)	Acquisition Retention Units Revenue

obsessed with lagging financial outputs (revenue and profits) over what's driving them.

In 2009, Jeff Bezos wrote in Amazon's letter to shareholders: "Senior leaders that are new to Amazon are often surprised by how little time we spend discussing actual financial results or debating projected financial outputs. To be clear, we take these financial outputs seriously, but we believe that focusing our energy on the controllable inputs to our business is the most effective way to maximize financial outputs over time."[8]

This insight captures the essence of the input-output framework. When you make decisions on which part of the customer journey to invest in, between product and marketing across acquisition, engagement, and retention, you need an integrated way to measure and evaluate initiatives and results holistically. That's why it's important to define the right set of controllable inputs, so you can optimize what truly drives business growth.

Integrating Brand and Business Performance Metrics

A holistic data reporting structure isn't just strategic but essential. A customer-centric measurement can shift a team's mindset from "Why isn't our business growing?" to "Why aren't customers engaging with us?"

An integrated measurement and reporting approach helps connect brand metrics with traditional business KPIs and makes them tangible. From awareness and discovery to acquisition to retention, you can look at the entire customer journey and understand which parts are growing or declining, so you can leverage the insights from measurement to inform your strategies and actions. This holistic measurement approach tracks how functional collaboration across product, marketing, customer support, and other departments drives growth together. Your brand metrics can become a decision-making tool, so they should be reported alongside revenue metrics to provide a more comprehensive view of the business.

Integrated Performance Metrics

DIMENSION	KEY METRICS
Awareness	Aided and unaided awareness among target customer segments
Perceptions	2–3 most important perception metrics influencing purchase decisions and customer satisfaction
Top of the Funnel	Website traffic volume, traffic sources, organic vs. paid visitors
Acquisition	Total sign-ups, conversion rate
Engagement	Product usage, engagement metrics
Retention	Retention rate, churn rate
Financials	Revenue, revenue per customer, profit margin

At Amazon, I incorporated three to four key brand health metrics into weekly business reviews. It was an effective way to outline valuable input and output metrics and help cross-functional teams understand what needed to improve. It also provided a natural avenue to get people to understand how each marketing initiative was connected to drive the business. For instance, we might discover that our delivery speed was fast, but customers had no idea. You could point to the perception metrics and say, "Look, customers don't even know we're this fast! We need to fix that!" Or maybe the data showed people didn't realize our new subscription program could save them hundreds of dollars, which instantly gave us a clear goal for improvement.

When you have metrics like these presented right next to financial metrics, it becomes natural and tangible for the entire organization to connect the dots and think about how each initiative is contributing to business outcomes. Through such measurements, you can gain insights into the effectiveness of creative work in capturing customer attention, assessing the efficiency of marketing channels, or understanding the appeal of the product and brand. You can act on the insights your data reveals to iterate your business.

Qualitative and quantitative data should work together, but there will always be unquantifiable elements and decisions that require human judgment, intuition, and passion. Business isn't always black-and-white. Not everything fits neatly into a spreadsheet. As Tobias Lütke, CEO and founder of Shopify, said in an interview on *Lenny's Podcast*: "We believe that having metrics make the decisions for us is an abdication of our responsibility to be the leaders of the space . . . I think there needs to be more acceptance in businesses of unquantifiable things. One of the most powerful unquantifiable things in the world of business are fun and delight."[9] Lütke's insight reveals the complexity of running a business and the importance of intuition and emotion on top of all the data you look at every day.

Once you have integrated measurement in place, you can use the insights to continuously optimize your strategy. The holistic data reveals how brand equity, campaign performance, and business results connect, helping you make decisions about where to invest and what to test next.

Embracing Experimentation and Iteration

Long-term thinking and an experimentation culture are not in conflict—you can do both. Today's digital tools make it easier than ever to test the effectiveness and viability of long-term strategies before fully committing to them.

Nike's $28 billion lesson teaches us that even in a data-driven world, you must experiment to validate assumptions. Nike's digital transformation strategy looked compelling—more data-driven, more direct relationships with customers, more streamlined operations. But by making sweeping changes to its business model, the company neglected what truly mattered to Nike's customers.

In his book *The Lean Startup*, Eric Ries advocates for rapid experimentation and iteration in product development. He emphasizes a build-measure-learn feedback loop, which encourages companies to quickly develop minimum viable products, test them with customers, and use the learnings to inform further development. This continuous test–learn–scale approach can be equally valuable in marketing and brand building. You can leverage real-time customer feedback from product experience, social advertising data, paid search results, social comments, customer service calls, and PR channel executions to inform your strategies. These insights and market signals from short-term executions can help refine targeting, channels, storytelling, and investment efficiency, as you refine your long-term strategies.

Incrementality Test

Many companies allocate a percentage of their revenue for marketing and advertising to drive demand. One major debate is often about paid media—what's the best percentage to invest in middle- to upper-funnel channels such as YouTube pre-roll and TV, and lower-funnel channels such as paid search? Marketing effectiveness, budget allocation, spending level, and ROI often fuel the debate. The good news is that you can test them.

I recently spoke with a friend who works for a major streaming service company in New York about shifts in media investment across the industry. Anecdotally, she's seen companies move from a 70/30 split to a more balanced 50/50 allocation between lower-funnel and mid- to upper-funnel efforts. This shift is important because it signals a broader recognition of the value of long-term brand building and the need to move beyond direct response campaigns focused solely on immediate conversions.

If your team has a similar challenge and can't decide the optimal allocation, you can run incrementality tests to understand the impact of improved awareness and perceptions on generating new, ongoing demand and driving continued growth. For example, you can test paid advertising on a small scale or in a limited set of markets for six months, learn from the results through data analysis and customer feedback, and then scale the most successful approaches.

You can also experiment on a tactical level. For example, my friend's media team stopped buying high-intent brand words for paid search. When customers searched these key branded words, they were high-intent customers. They already knew the brand and why it mattered to them. You can capture this organic search traffic to generate sales without paying for those search results. Through a test run, her team found no drop in performance after turning off high-intent keywords. This might sound like a small tactic, but small steps add up to meaningful changes.

The experimental and iterative approach helps you demonstrate quick wins and ensures confidence in the strategy. You can be committed to your long-term vision while staying flexible with short-term execution and tactics. As you grow, your customers and market environments are evolving, and their relationships with your products and business need to evolve to stay relevant. This evolution often signals the need for new product offerings, expanded services, or different engagement approaches.

Evolving with Customers to Stay Relevant

Many companies start with one product to enter the market and gain initial customers, then launch multiple offerings as they grow, as we saw with Rivian, Ring, and Robinhood, among many others. As you scale and manage complex initiatives that span across various product lines, sales channels, and different time frames, it's important to stay true to your core customers while expanding to meet their evolving needs.

The process of "measure, iterate, and scale" helps you turn insights into strategic actions to refine product, marketing, and business strategies. These insights can inform decisions on what new products and services to develop next, how to engage with customers, and how to support their growing needs.

For instance, Robinhood's product evolution mirrors their customers' journey. What started as a mobile trading platform for millennials has evolved to offer eight different financial services. As customers get older, their financial needs evolve and they want retirement accounts, high-interest savings, and asset management tools. Robinhood grew alongside them and expanded their service offerings.

Nike demonstrates this evolution at scale through its category strategy. A six-year-old might first encounter Nike through a basketball camp, but as they grow older, they need tennis shoes and fitness apps for tracking activities. Nike's basketball, soccer, running, tennis, and youth programs cater to these different customer needs while maintaining a consistent brand promise.

As you iterate on your strategies, remember that brand building is ultimately about building meaningful relationships with customers. To do this, you need to grow and evolve with your core customers. This brings us back to where every journey begins: define customer needs, develop better products and experiences, and engage with them over time. For product, the goal must be to solve customer problems. For brand, the goal must be to create happy and satisfied customers who trust you. Everything contributes to the ultimate business goal: creating and maintaining strong customer relationships.

Chapter Summary

Focus on what truly matters to customers. In a data-driven world, it's easy to fall into the trap of focusing on what's simple to measure. While digital transformation and measurement are important, they must be grounded in what customers truly value about your brand and products.

Brand equity is the engine of sustainable growth. Awareness, perceptions, and satisfaction provide key signals about potential and existing customers. These long-term metrics help you understand your position in the market and reveal barriers and motivators behind customer acquisition and retention.

Overcome attribution silos through integrated campaign measurement. Full-funnel metrics, varying time frames, and shifting business prioritization make measurement genuinely complex. But success requires considering the full customer journey and connecting upper-funnel metrics with lower-funnel results to measure the complete impact.

Focus on your controllable inputs, not just outputs. The input-output framework emphasizes what you can directly influence, such as value proposition, customer targeting, product experience, and creative work, rather than obsessing over lagging financial results.

Integrate brand metrics with business performance. By incorporating brand health metrics alongside traditional business KPIs in a holistic reporting structure, you create greater organizational alignment and visibility into what drives results.

Use experimentation to validate long-term strategies. Stay firm in your long-term vision but flexible in execution. Through incrementality tests and controlled experiments, you can demonstrate quick wins, build confidence, and refine your approach based on customer feedback.

Evolve with customers to stay relevant. As your customers mature, their needs change. Your business must evolve alongside them, staying true to your promises while adapting how you serve customers over time.

PART IV

Support a Brand-Power Culture

The final part of our journey is about your workplace culture.

Culture is the foundation that shapes how your company operates, how you hire and retain talent, and how your teams work together to achieve your mission and deliver for customers. Your brand power comes from within, so your culture must embody your promises to customers, employees, investors, and partners.

Your products are also a direct reflection of your culture. What happens inside your organization ultimately manifests in what customers experience. The values you prioritize, how you make decisions, and how your teams collaborate will all show up in the products and services you deliver.

Your culture shapes your brand identity, drives product innovation, and defines how your company grows and sustains success in the long term. It's hard to overstate its importance. So let's explore how to build a customer-centric culture that will fuel your brand power from within.

Chapter 9

Cultivate a Culture That Powers Brand Success

From Organizational Silos to Customer-Centric Culture

To truly innovate and win customers, your brand vision must be embedded in your workplace culture, with the entire organization living and breathing it every day. A customer-centric culture is the foundation of brand power, but maintaining this focus becomes increasingly challenging as you grow and scale.

The organizational design, structure, and processes are fundamental to turning vision into reality. This chapter explores how to cultivate a culture that always puts customers at the center and empowers teams to deliver great customer experiences together, even as organizational complexity increases.

How Figma's Culture Transformed Design Collaboration

In 2012, Dylan Field and Evan Wallace, two students at Brown University, discovered the potential of web browser and cloud-based applications to facilitate collaboration and design sharing. They started building a web-based product design tool called Figma, which became the first design tool that combined the accessibility of the web with the functionality of a native app. In 2016, Figma's first paid product officially launched with the mission to make design accessible to anyone.

In the past, design tools had been made and developed mostly for individual designers, not for collaboration. For instance, Sketch and Adobe were created for professionals and required local installation on each computer. The capabilities of these tools discouraged non-designers from engaging with them. But now, Figma fundamentally changed design and made it accessible to everyone. The startup promised that the future of design wasn't just about better tools, but about transforming how teams work together to deliver better experiences.

Figma's early innovations centered around three key areas:

- It started as an online and cloud-based software that enabled access from anywhere without local installation.
- It offered simultaneous multiple editing, allowing teams to work together on the design of products and experiences in the same file.
- It enabled users to provide inputs and feedback in one centralized place.

With these new features, product and experience design suddenly evolved from one person's job to a collaboration among product managers, engineers, marketers, and other non-designers. Figma's approach moved design files from fragmented places to a single destination. These may seem like technical changes, but their impact was profoundly cultural. Figma helped shift design thinking from a designer's skill to a widely shared mindset.

In the early days, the design community rejected Figma's approach to protect their independence. As CEO Dylan Field wrote on the company blog: "Initially I didn't understand the negative reactions to Figma's closed beta launch . . . Now I understand that the power of the browser lies in the broader cultural change it delivers—and this change can be scary. The browser is natively multiplayer. It forces a mindset shift on access. It strips away the need for expensive hardware. And it pushes us to embrace working together."[1] Today, Figma is a serious player in the enterprise software space with over 13 million monthly active users, and it became a public company on July 31, 2025.[2] It serves the creative and engineering community from design all the way to code and production, with products such as Figma Design, FigJam, Dev Mode, and others.

But perhaps most interesting is *how* Figma achieved this transformation in its field.

A Culture to Make It Happen

Figma's rise is inseparable from its culture and values. The web-first approach wasn't just a technical shift but reflected their belief in openness and access. To make design accessible, Figma first had to be open and collaborative. Otherwise, how could they expect it to enable those values for their customers?

In 2021, 96% of employees at Figma said the company was a great place to work,[3] compared to 57% at a typical US company. Dave Vega, former Director of IT, explained why: "What I love about Figma is that we are a people-centric company with a people-centric culture. This allows me to work on finding innovative tools that help people do their jobs in a way that isn't 'big brother,' but something our employees will truly enjoy."[4]

Figma's openness reflects its internal process. The company uses its own product FigJam to brainstorm and define ideas. FigJam's interactive feedback mechanism allows everyone to see priorities, and then decisions and feedback are centralized and accessible to employees.

For example, the biggest job function at Figma is engineering. By running engineering critiques,[5] an informal review before the final technical review, it allows engineers to share early thinking and solicit feedback from as many as 200 people across the entire organization. The engineering crit became so valuable that, over time, it has become a core step to develop and review technical designs within the company.

The company's ground-up innovation was critical to its success. For example, a Figma feature called Multi-edit, which allows users to simultaneously edit objects and text layers across multiple frames, was proposed by team members during the "Maker Weeks" when every Figma employee can work on anything as long as it benefits the company. Even though this design feature was a simple improvement, it made the design process flexible and easy to use.

In December 2023, Adobe's $20 billion acquisition of Figma fell through due to regulatory concerns in Europe. Many employees had joined Figma while the merger was in progress, assuming that they would become Adobe employees. When the acquisition fell through, some were disappointed. Dylan Field had frank conversations about the culture of Figma at an all-hands meeting and offered employees the option to voluntarily leave. Despite the failed deal, 96% of its 1,400 employees stayed[6] and only 4% took a buyout offer. This demonstrates the strength of Figma's strong culture and resilience.

The company's open and collaborative culture manifests in its product. Figma not only provides tools for companies to design products and services but transforms how teams approach creating great customer experiences. Figma's impact lies in its ability to shape other cultures and make a meaningful impact on the entire product building community:

designers, product managers, engineers, and everyone else who contributes to building great experiences.

How Your Culture Shapes Your Products

What exactly is culture? As Daniel Coyle, author of *The Culture Code*, put it, "Culture is a set of living relationships working toward a shared goal. It's not something you are. It's something you do."[7] Your culture is the invisible infrastructure that supports how people from different job functions collaborate to serve customers.

This isn't just theory—it has measurable business impact. A Harvard study[8] researched more than 200 companies and discovered that a strong culture increases net income 765% over 10 years. Your product is a direct reflection of your culture; whatever happens inside your organization shows up in your products and services. The values you prioritize, how you make decisions, and how your teams collaborate will all show up in the products and services you deliver. Your brand power comes from within, and your culture is that foundation.

Culture manifests differently across companies, but always shapes the customer experience. Whenever I'm asked about my experience working with Apple, I like to joke that "We used a ruler to measure our keynote presentations!" Every detail matters. Craft matters. That's Apple's culture of high expectations, which extends to everyone who touches whatever Apple delivers to customers. Even as it grew to become one of the largest companies in the world, the high bar for detail and craftsmanship never got lower. That exceptional brand consistency required a very strong culture to maintain.

When I first started working at Amazon, I was shocked by its peculiar attitude toward meetings. Every meeting starts with everyone reading a six-page memo, drafted by the meeting organizer, in silence. This unique practice helps drive leadership alignment and good decisions. It was a cultural shock because I had come from Apple's approach, which was obsessed with simplifying everything. Folks there would have hated the whole idea of the six-page memos. But, over time, I came to appreciate Amazon's document culture, along with its "working backwards" mechanism, data discipline, operational rigor, and all the other cultural elements that make Amazon successful. It's a different kind of excellence, not necessarily better or worse than Apple's.

Building and maintaining a customer-centric culture through growth and technological changes requires more intentional design. What we discussed in the Six Building Blocks section may feel like common sense. You may have been nodding along, wondering why everyone doesn't do these things already. The answer is that building this kind of culture is extremely difficult! I see six essential cultural strategies that can help you transform your vision into reality and keep your focus on customers:

- Anchoring your culture around your North Star
- Investing in the humans behind the work
- Embedding marketing at the core of innovation
- Designing organizations to prevent fragmentation
- Sustaining culture through growth and change
- Leading through AI without losing customer focus

Let's go deeper on each of them.

Anchoring Your Culture Around Your North Star

As we saw in chapter 3, your North Star establishes the fundamental direction and purpose of the company, guides your brand and product development, shapes your culture, and informs everything you do and the employees you hire. When a strong company culture is anchored to your North Star values, it empowers teams to consistently deliver for customers.

The culture comes from the mission: why you do what you do. Start-ups often begin with strong cultures that obsess about a core problem to solve, about trying to shake up the status quo and deliver excellent customer experiences. In the early days, it's easier to stick to the vision even in challenging times.

Every company we've focused on in this book—including Airbnb, Rivian, Ring, PillPack, Notion, Robinhood, and others—has a culture that laser focuses on solving well-defined problems for their customers.

According to Daniel Coyle, purpose is one of three key elements of successful cultures, alongside building safety and sharing vulnerability. A company's mission creates a sense of belonging and purpose that connects all employees. This cultural foundation is what allowed Airbnb to revolutionize the way we travel, Rivian to become the most loved car brand, Ring to transform home security, PillPack to reimagine online pharmacy and medication management, Notion to redefine workspace collaboration, and Robinhood to democratize finance.

Let's look at how this works in practice. Take Figma, for example. Their success wasn't built on technology alone, but on core values that shaped both their internal culture and their products: "Build community," "Run with it," "Love your craft," "Grow as you go," and "Play." These values directly supported their mission to make design accessible to everyone. Each value reinforced their commitment to openness, collaboration, and democratizing design, which became the foundation for every product decision and cultural practice at the company.

Notion demonstrates another powerful way to anchor culture around a North Star. The company named its conference rooms after timeless tools to remind and inspire employees about the value of product innovation and craftsmanship. One room is called "iPhone," another

"Toshiba rice cooker," and another "Sony transistor radio." These inventions changed countless lives and lasted for decades. According to founder Ivan Zhao, the intention was to inspire the team to build something lasting, even in an era when companies move fast and technology changes constantly. This simple office design choice reinforces their commitment to creating tools that last.

Every successful product and brand is built by people who create and embody their workplace culture. While culture shapes behavior, it's the leaders of the organization who create culture in the first place. If work is ultimately about people collaborating toward a shared goal, a culture that shapes people is the true foundation for brand power.

Investing in the Humans Behind the Work

One of the greatest brands I admire is Nike. Its founder, Phil Knight, shared a powerful insight in his memoir, *Shoe Dog*: "It's never just business. It never will be. If it ever does become just business, that will mean that business is very bad."[9]

That line has stayed with me. In the day-to-day focus on deliverables and outcomes, it's easy to forget that business is fundamentally about people—the customers you serve, the team you build with, and the partners you work with. Behind every great product and company is a network of human relationships and shared vision. Culture is the emotional connection people have with each other and it's what drives business forward. Business is nothing if it's not about people.

Jim Collins, in his famous book *Good to Great*, captured an enduring business truth: "The right people will do the right things and deliver the best results they're capable of, regardless of the incentive system."[10] Conversely, the cost of hiring the wrong people or losing good people is high. Harvard Business School professor Tom Eisenmann,[11] who researches why startups fail, has written that "a broad set of stakeholders, including employees, strategic partners, and investors, all can play a role in a venture's downfall."

When I spoke with Eric Grunbaum, VP of Creative at Airbnb, about the importance of company culture, he emphasized the vital role of leadership in fostering an environment where teams feel a true sense of belonging and can do their best work. He shared a powerful insight: "Creative ideas are accidents. We can't control when they will happen, and there's no machine that cranks them out on demand. But what we can do is cultivate the conditions that allow creativity to flourish, and make it more likely an accident will happen." Before joining Airbnb to lead its global brand and creative efforts, Grunbaum spent over a decade meeting with Steve Jobs every week to review Apple's global advertising campaigns. So he knows a few things about culture.

When employees feel safe to take risks, they explore bold, unexpected ideas. But without that trust, they may only pursue what's safe and predictable. That's why culture is the foundation of innovation and creativity.

Remember the PillPack story from chapter 5? PillPack also demonstrates how deeply internal culture shapes customer experience. For instance, the company developed an empathy exercise[12] that immersed every new employee in the daily challenges faced by their customers. Employees sorted pills while wearing oversized gloves to simulate arthritis and thick glasses to mimic poor eyesight, experiencing firsthand the physical and emotional struggles of managing those chronic conditions. This empathy training was critical for every team member to understand why their work mattered.

A focus on people and culture has become increasingly critical for business success. A Glassdoor survey revealed that 77% of respondents consider culture before applying for a job at any company,[13] and 56% of respondents consider company culture more important than salary for job satisfaction. Yet, according to the US Bureau of Labor, the average employee tenure decreased to 3.9 years in 2024,[14] with young people aged 25 to 34 staying only 2.7 years on average. When employees leave, the continuity of work suffers, institutional knowledge is lost, and the culture itself can begin to erode.

But building a strong culture isn't about just creating a "happy" workplace.[15] People from the most successful cultures are oriented more around solving hard problems together than achieving happiness. This mindset becomes especially important for cross-functional collaboration.

Embedding Marketing at the Core of Innovation

At many companies, marketing is seen as a service function, pumping creative assets and managing campaigns. Some even see marketing as a cost center. Over the years, many industry friends have shared the concern that the marketing team doesn't have enough influence or "a seat at the table" when important strategic decisions are made.

If you lead marketing, you may wonder how you can elevate the marketing team's role in the organization, especially in the product development process. The *Brand Power Built In* approach brings the functions of marketing and innovation together and fundamentally changes how companies view marketing—not as a creative or execution function, but as a strategic partner in value creation. Marketing is the voice of the customer and can help drive customer-centric decisions across the organization. It should have a voice in product development from day one.

This means inviting marketing leaders to initial product ideation meetings, contributing to product requirement documents, and establishing regular cross-functional brainstorming sessions. Such inclusion will ensure that brand and go-to-market considerations are baked into the product from the start, leading to better products and more successful launches.

Companies that position marketing at the core of innovation tend to succeed. Think about how Rivian focused on a niche market to become the most loved automobile brand in America. Think about the early days of Ring and how strategic branding launched Ring to success. Think about PillPack's unified approach to deliver value to customers. They all reflect a cultural understanding that marketing and innovation are inseparable as both functions drive a business forward.

Peter Drucker, the renowned management thinker, once said, "Because the purpose of business is to create a customer, the business enterprise has two—and only two—basic functions: marketing and innovation. Marketing and innovation produce results: all the rest are costs."[16] This might be an oversimplified view of organizational functions, but it speaks to the cultural centrality of marketing and innovation.

Designing Organizations to Prevent Fragmentation

Organizational fragmentation is a common challenge when you scale and launch many products and services, and the business gets more complex. Remember the "clean up the notifications!" example from chapter 7? Each team at that company only oversaw a part of the customer journey, and no one thought it was their job to optimize the entire customer experience. That was a typical example of how a fragmented organizational structure leads to poor customer experiences and lower satisfaction. What happens inside your organization directly impacts customer experience.

Navigating organizational challenges is probably the most difficult part of building your brand power. Even the most talented employees can be hindered by ineffective organizational structures. The daily pressures of business pull you away from customer focus, and it's easy to end up *internally* focused on your own needs and goals. It's easy to focus on what the company wants and what various other stakeholders want, rather than what customers want.

When you expand and introduce more products, teams, and business units, the magnitude of the challenge and the complexity increase exponentially with size and scale. You need to manage more people, more processes, more cross-functional collaboration, and more leadership alignment, often with different goals and competing priorities. When I left Amazon in 2024, the company had more than 1.5 million global employees across hundreds of business units. Just imagine the challenge of managing cross-functional work within such a giant company.

How can you foster cross-functional collaboration to deliver cohesive customer experiences? The design of the organizational structure and processes is fundamental to ensure that different parts of the customer journey work together.

Clear Roles and Responsibilities

Organizational fragmentation can manifest in many ways. I once got into an intense debate with another leader about the division of responsibilities between brand marketing and product marketing. This leader's business unit represented both a product and a brand, and it was still in the early days of growth. So, in a sense, its product marketing and brand marketing were synonymous—improving one would automatically improve the other.

But this leader wanted to grow her head count to mirror the product department's expansion and growth. So, her solution was to cut the work into pieces to create new responsibilities for new roles. She tried to claim that "brand marketing" was responsible for awareness and perception metrics, while "product marketing" was responsible for product adoption and growth. As a result, I saw teamwork get torn apart, with two teams working on essentially the same goals, without clear ownership or accountability. I realized that people misunderstood what product marketing and brand marketing truly mean, and this misunderstanding led to an organizational design that created fragmentation and ultimately got in the way of solving real customer problems.

For example, to improve awareness and perceptions, a brand marketing team must develop positioning, messaging, targeting, channels, storytelling, and creative work—the essential elements for any campaign to get people's attention and inspire them to take action. But if brand marketing already does this work to produce a campaign, then what work is left for product marketing?

Conversely, for product marketing to execute, the product marketing team must produce positioning, messaging, targeting, channels, and creative work. How much room does that leave for the brand marketing team to come up with growth-driving brand campaigns? What's worse, if improved awareness and perceptions don't directly translate to growth in traffic, conversion, revenue, or other key metrics, what's the point of doing brand marketing?

Such day-to-day executional challenges happen all the time when roles and responsibilities are not clear. Even when leaders don't intentionally put two teams to work on the same goals, they may accidentally be setting up conflicts and overlap. The result is guaranteed frustration and wasted effort internally, plus lost trust among external stakeholders who don't know who their partners are. It's a terrible combination.

Even though this might sound like a minor example, the damage to team morale and workplace culture can be significant. If team members become complacent, they will avoid accountability. And when the business fails to grow or meet other objectives, it's hard or impossible to identify why. Ownership spread across multiple teams ultimately means no ownership at all.

The Need for Clear Accountability

This organizational fragmentation often gets in the way of producing great work because it lacks a single threaded owner for clear accountability.

I learned this from Apple's senior vice president of engineering, Dan Riccio, who shared an important lesson after working with Steve Jobs and Tim Cook for 26 years. His view was that a "matrix" organization of multiple owners lacks accountability. When things go wrong, multiple managers are more likely to make excuses rather than take responsibility for fixing problems.

Some companies are functionally organized while others are matrix organizations, with general managers overseeing different business units. Regardless of how you structure or restructure your business, make sure all lines of responsibility have clear RACI—an acronym for Responsible, Accountable, Consulted, and Informed. For each function, or workstream, who is ultimately responsible? Who needs to be consulted? Who merely needs to be informed?

I find the RACI model quite liberating as it provides a clear structure to help cross-functional teams understand their roles and expectations, which makes their work clearer and more efficient.

I once asked my mentor, a VP responsible for a 300-person business unit, about the most effective organizational designs. Throughout her career, she witnessed numerous reorganizations, and some she personally implemented. She explained that some reorgs were driven by expanding a leader's scope as a reward for a bigger job title. Others were driven by consolidating resources to reduce cost. But the most effective reorganizations focused on solving customer problems and optimizing the customer journey. That was the best way to get best results from leaders and employees alike.

To build lasting brand power, you must design a customer-centric organizational structure, encouraging clear ownership and accountability with the priority of making things better for customers, first and foremost.

Sustaining Culture Through Growth and Change

Startups often begin with a strong, mission-driven culture that fuels innovation and early success. But as companies scale, many struggle to sustain and strengthen their culture. It takes leadership and intentional design to lead through change and growth without losing purpose and customer focus.

When Culture Goes Wrong

When you get the culture wrong, the costs can be devastating. For instance, when Uber had a culture crisis in 2017, it almost destroyed the company. Obsessing over disruption and hyper-growth made Uber successful in the early days, but it also fostered a "win at all costs" environment in which colleagues turned against each other as competitors. Then the crisis of negative publicity about the company's practices ultimately led to the resignation of cofounder and CEO Travis Kalanick, along with significant employee turnover. Worse, the culture crisis led to a viral #DeleteUber social media movement, which made it hard to recruit new drivers and seriously impacted business growth.

Under new leadership, Uber adjusted its cultural values and pivoted to focus on building a culture that would reward teamwork and long-term employee commitment. As new CEO Dara Khosrowshahi wrote in a blog post, "It's also clear that the culture and approach that got Uber where it is today is not what will get us to the next level. As we move from an era of growth at all costs to one of responsible growth, our culture needs to evolve."[17] As part of the changes, Uber removed some values from its official list. For example, it no longer encouraged "toe-stepping"—a practice meant to encourage the honest sharing of ideas but had often been used as "an excuse for being an asshole."[18]

As you can see from this cautionary tale, a company's culture and people are the foundation for long-term success and brand strength. What worked in the early stages of a startup may not work for the scaling and growth phases of a more mature company.

Maintaining Growth and Change

In fact, organizational culture is so powerful that even the same products and people can yield different results when placed in different cultural environments. I witnessed this firsthand when Ring was acquired by Amazon. Ring had built a strong, distinctive culture that was key to its success. However, the integration into the Amazon system required significant time and thoughtful navigation, a common challenge in major acquisitions.

The integration process revealed how deeply culture influences every aspect of operations, from major areas such as decision-making processes, doc writing mechanisms, and customer engagement philosophies. They also showed up in seemingly small areas such as preferred productivity and collaboration tools. For example, Ring at the time used Google Workspace, while Amazon relied on its own collaboration tools. What might seem like a simple technology choice became a symbol of stark cultural differences, with teams working to bridge the tension between a startup's scrappy roots and a mature, established process.

That experience made me realize the deep impact of organizational culture. It's not just about shared values; it's also about how the culture guides people to make decisions, collaborate, and deliver for customers.

What happens inside the organization manifests in your products and customer experiences.

Leading Through AI Without Losing Customer Focus

I'm writing this at the beginning of an AI transformation. From developing code to creating UI/UX to generating marketing content, virtually every tech company is embracing the promise of AI.

With these tools becoming an increasingly significant part of your day-to-day work, you may wonder what it will take for your organizational culture to evolve and endure through this transformation. How can you still maintain customer focus? How can you embrace the promises of AI while raising the bar for quality and customer experiences?

Balancing Efficiency with Customer Value

The integration of AI technologies can potentially further amplify organizational fragmentation and impact customer experiences. Your work may become even more complex when different job functions and teams adopt AI. Your websites, content, product design, and experiences can be (at least partially) generated by AI, but in the process it's easy to expose complexity to customers and create customer confusion.

Being efficient doesn't necessarily translate to better quality of work or more effectiveness. Cutting costs alone is not enough for a business to succeed in the long term. You need to consistently create value and connect with customers in meaningful ways.

One problem to watch out for is AI-powered misinformation. According to a survey[19] conducted by Frontify among 500 chief marketing officers (CMOs) in the United States and the United Kingdom, 74% of CMOs are concerned about "fake brand partnership" as their biggest worry. With brand spoofs and unsolicited collaborations on the rise, CMOs see an urgent need to protect their brands from new risks.

When you use AI chatbots to replace human customer service, it's more efficient and reduces your operational cost. But when not designed with customer needs in mind, the change can damage the customer experience and erode customer trust. According to a survey, 81% of people reported that they would prefer to wait a minute or more for support from a live person rather than interacting immediately with an AI assistant.[20] You need to seek an optimal balance between efficiency and customer needs—not easy.

Brand Is Your Greatest Asset

A strong brand has always been about differentiation and trust, and that still applies to digital experiences that you hope will connect with customers in a meaningful way.

As Dylan Field says about the importance of design in digital experiences: "Design will be the way that you differentiate that software and make it something that people want to use and your business wins. Or if you're not strong enough in design, that's why you might lose."[21]

While AI capabilities may empower you to create things faster and cheaper, you can't forget the fundamentals: Technology is merely a tool to serve the higher purpose of winning and serving customers. You can never forget why you do what you do. Without the *why* to guide the design of your entire customer experience, it's easy to confuse efficiency with quality and customer satisfaction.

Nothing is possible without the people who work behind the scenes to bring value, delight, and surprise to customers. The decisions you make along the way, the subtleties you discover, the trade-offs you consider, the workplace culture you build—all of these human elements remain irreplaceable.

In the end, it's brand power, human connection, and emotional resonance that help you stand out, attract customers, and really win the hearts and minds of customers.

──────────────── **Chapter Summary** ────────────────

Your culture directly shapes your products. Your product is a direct reflection of your culture, which is the foundation of enduring brand power. Culture shapes your brand identity, supports your product innovation, and impacts customer experience.

Anchor your culture around your North Star. Your North Star values guide your brand and product development, shape your culture, and inform everything you do and who you hire. The strongest brands stay true to their mission even as they scale.

Invest in the humans behind the work. Every successful product and brand is built by people who create and embody the culture, making them the true foundation for lasting brands.

Embed marketing at the core of innovation. Marketing and innovation are inseparable as they combine to form the core of business growth. Companies that position marketing at the core of innovation outperform those that maintain silos.

Design organizations to prevent fragmentation. As you scale and launch many products and services, silos and fragmentation will tend to emerge. Success requires designing structures around customer needs, establishing clear accountability, and putting the customer at the heart of all decisions.

Sustain your culture through growth and change. It takes leadership and intentional design to lead through change and growth without losing its purpose and customer focus. When a culture goes wrong, the costs are devastating to both reputation and business results.

Lead through AI without losing customer focus. In the age of AI, brand is your greatest asset. Success comes from maintaining focus on meaningful customer experiences rather than just efficiency—fostering trust and human connection that technology alone can't provide.

Putting It All Together

Technology Without Purpose: The Story of Humane Ai Pin

On February 18, 2025, Humane announced to its users that it would soon permanently discontinue support for the Ai Pin, a small, chest-worn, voice-powered wearable device. Shortly thereafter, HP acquired Humane for $116 million.[1]

The Humane Ai Pin had been announced in November 2023 with a vision to reduce our reliance on screens. It was widely described as one of the most promising AI-driven devices. Yet less than two years after its announcement and a year after its 2024 launch, it was shut down. Why did it fail so completely and so quickly?

With a device starting from $699 and an additional $24 per month subscription for a wireless service fee, Humane promised to replace your smartphone for several common use cases. As their marketing copy put it, "Whether you're making calls, sending messages, seeking answers, capturing moments, taking notes, or managing your digital world, Ai Pin acts as your assistant and second brain, allowing you to be present and in flow."[2] Humane had a well-designed website, and the device had an elevated, premium aesthetic.

But when early adopters tried the product in 2024, they mostly hated it. *The Verge* reported that thousands of units were returned, quickly outpacing early sales. People reported a significant gap between what was shown in demo videos and the actual customer experience—even though the founders had stated publicly that the Ai Pin was still a "version 1.0" with improvements certain to follow. In the wake of the failed launch, critics pointed to the unreadiness of the product to go to market. Common complaints included too few features, poor battery life, problems with overheating, glitchy AI performance, and more.

But I see this not primarily as a case of failed technology so much as a failed brand and go-to-market strategy. As a famous advertising saying goes, "Nothing kills a bad product faster than great advertising." We should give Humane credit for attempting to push the boundaries of wearable AI, an innovative new product category. Innovation is inevitably risky and always involves failure. Humane's big mistake wasn't trying to challenge the status quo—it was ignoring several core principles that I've stressed throughout this book.

- Instead of focusing on their proprietary technology, companies should obsess over a core customer problem to solve. The Ai Pin had very cool new tech, but its customer benefits and purpose were never clearly defined. Those early adopters found that it failed to serve any true need.

- Customer connection and trust must be embedded into the product experience from the start. The poor product experience should have been identified and fixed before it went to market. You can't launch a product and then retroactively give it purpose, heart, and a trustworthy brand. Customer connection and trust need to be priorities very early on, when the product is still in development.

- Launching too big and scaling too quickly is dangerous. It's far better to move slowly and keep market testing limited until you're sure you've achieved product-market fit. Whenever there's a misalignment between a company's ambitious vision, an inadequate customer experience, and a "get big fast" go-to-market strategy, there's no margin for error—and little opportunity to iterate and improve. The result tends to be overpromising and underdelivering, negative PR, massive customer disappointment, broken trust, and terrible word of mouth.

Your Brand Is the Heart of Your Products

You see, building a brand isn't about creating a fancy website, an inspirational story, or bold marketing and advertising campaigns. Those are all important elements, but for a company to win the hearts and minds of customers, the product first needs to *deliver*.

First and foremost, start by obsessing over a core customer problem, and then develop a product to solve this customer need. Your story, your products, and features should work together to deliver functional and emotional benefits that solve that need. It's unlikely to work if you start with an innovation and then go looking for a use case.

Ask the fundamental questions: Why does your company exist? Who are your ideal customers? Why should those customers care about you? These questions may seem too fluffy and obvious to spend time on, but they're not. These are actually the *hard* questions—and the answers will be the invisible threads that drive your brand and product strategy.

Humane Ai Pin failed to embed brand thinking into product development from the start. Despite having impressive technology and high-quality marketing content, the product didn't deliver the functional or emotional benefits that it promised. As a result, customers quickly lost trust—and not even the coolest tech or the most lavish marketing can replace customer trust, or the purpose of a product. Those elements form the foundation of your story and product experience.

Humane's mistakes may seem obvious in retrospect, but I can imagine how difficult and painful their journey was to create and launch a pioneering product. Their failure reminds me of the email I received from a founder friend, which I shared in chapter 2. As you may recall, my friend wrote:

> When you try to build something, there are so many difficult and time-consuming tasks that you have to make trade-offs. Sometimes, things that feel ethereal or less tangible, like your story and your reason for being, don't make it through. Those are the harder things to communicate and those are the more difficult things to have other people understand. It's very easy for someone to say, "We need to build a website and it needs to do these things," but it's hard to say, "It has to have the story and it has to have this feeling so our customers will feel a certain way" . . . I've had to trade them for just very short-term progress. And I can now understand and see how it's a slippery slope that you can wake up one day and you just didn't get what you meant to get.

If you're trying to create something new, it's important to constantly remind yourself that customer connection and trust need to be baked into the product experience. You need to press yourself to see if your product experience truly delivers that promise. Bring brand thinking upstream as early as possible during product development, to assess marketability, reduce risks, and improve your odds of success. The sooner you integrate these principles, the more likely that your product will achieve a true customer connection.

Another key lesson: Launching a product requires deep customer understanding and a well-integrated brand strategy and go-to-market strategy. You should aspire to a sequenced launch to establish product-market fit *before* investing in mass-marketing and scaling efforts. If Humane had taken a more measured approach, testing with early adopters and refining before trying to scale, the outcome might have been very different. It's important to balance your bold vision with the patience to fully address negative customer experiences before you go wide. The alternative is a potentially fatal breach of customer trust.

Like the internet and smartphones in past decades, AI is now fundamentally changing customer relationships: how we search and discover, how we interact with machines, and how we connect with one another. Yet our fundamental human desires remain constant. What makes us human and what makes our hearts sing will not change. The most enduring brands aren't just built on algorithms or automation—they're built on human connection and trust.

Your core mission is serving customer needs and earning trust so people will want to tell their family, friends, colleagues, and communities about your products, not just immediately but year after year.

Build Your Brand Power

Thank you for sticking with me throughout this book. Together, we've explored how to build brand power from the ground up and why it matters more than ever. Whether you're a product builder, marketer, or business leader, you have an opportunity to shape your product design and customer experience from the start, not when it's too late to make a real impact.

You have an opportunity to integrate long-term brand-building efforts with the product development process, not treat them as two separate activities.

You have an opportunity to invest in meaningful customer relationships across the entire journey and life cycle, not just one part of the journey.

You have an opportunity to pull customers in gradually and organically, rather than trying to scale too quickly through the brute force of big marketing budgets.

You have an opportunity to make the leap from good to great.

Perhaps most importantly, you have an opportunity to build an internal culture that will empower people to naturally build great products and customer experiences together.

Brand is a unifying force that compounds over time. The work you do in the building phase creates a strong foundation for your go-to-market phase. Once you establish product-market fit and gradually solidify your market position, you can gain a strong foundation for scaling.

There's no silver bullet or magic wand for any of this. Any strong brand is the culmination of smart strategy, intentional sequencing of key actions, and hard work.

As a former long-distance runner, I learned that there's no real finish line. Long-term success comes from continuous training and improvements. Building a great product, brand, and business requires the same mindset of a continuous journey. Sooner or later, it's likely that your customers will change, your competitors will change, and your market environments will change. But if you remain flexible and adaptive to the dynamic nature of running a business, you can change along with them.

From Apple to Amazon, from Airbnb to Robinhood, we've seen that brand building isn't a downstream marketing activity distinct from product development. It is a powerful value creation tool that integrates with product vision and business strategy right from the start.

Every business is different, but I hope that this book has inspired you to rethink your product, your brand, your business, and your culture—all with the goal of bringing your people together to deliver great customer experiences and build lasting customer relationships.

That's how tech products *really* win hearts and minds.

Glossary of Terms

I believe in the power of having a shared vocabulary to drive clarity and understanding, so I put together this basic glossary. It doesn't attempt to cover everything related to brand power, but I hope it clarifies some important terms that are often confused with each other.

What is **Brand Power Built In?**

Brand Power Built In is a repeatable framework for embedding brand power into a product's DNA from day one—not as an afterthought, but as a lasting competitive advantage.

> **Brand Power:** The strength of a brand to win customers' hearts and minds and drive both product success and business growth.
> **Built In:** An integrated approach to embedding brand power into a product's DNA from day one.

What is a brand?

People often use "brand" and related terms to mean very different things. Others may disagree, but here's how I define some key concepts.

Brand: Your customers' perception of and relationship with your company, products, and services.

Brand Identity: The tangible and intangible elements (logo, visual system, user interface, user experience, tone, and voice) that shape how your product is recognized and experienced.

Branding: A disciplined process of creating a differentiated identity in customers' minds.

Brand Campaign: A specific marketing activity designed to support overall business goals.

Brand Building: Activities across all touchpoints—from product experience to marketing campaigns—that establish and strengthen customer relationships.

Brand Equity: The measurable impact of the strength of your customers' perceptions and relationships on business outcomes.

What is the customer lifecycle?

Understanding how customers discover, engage with, and stay loyal to your product helps frame how and where brand power should be embedded.

Customer Lifecycle: The entire journey of a customer's relationship with a product, from initial awareness and discovery through adoption and long-term loyalty.

Customer Acquisition: The strategies and channels used to attract and convert new customers to your product.

Customer Engagement: The ongoing interactions that keep existing customers actively using and finding value in your product.

Customer Retention: The strategies employed to prevent customer churn and deepen existing relationships to extend their lifetime value.

Customer Loyalty: When satisfied customers become advocates and recommend your product to others.

NPS: Net Promoter Score, which is measured by asking customers a single question: "On a scale of 0–10, how likely are you to recommend our product or service to others?"

What is the marketing funnel?

The funnel typically refers to the customer journey from initial awareness to the purchase decision. It includes both upper-funnel and lower-funnel activities.

Upper-Funnel Marketing: Activities that focus on building awareness and positive perceptions about your product or service.

Lower-Funnel Marketing: Activities that drive specific customer actions, such as sign-ups, free trials, and purchases.

What strategic foundations are needed before launching a product or service?

A successful product launch requires clear strategic foundations. Here are the essential elements that often determine market success.

Positioning: A strategy that defines your brand and product's unique place in the marketplace and in the minds of your customers, serving as a North Star to guide organizational decisions.

Customer Segmentation: The process of dividing your total addressable market into distinct groups with similar characteristics to identify and prioritize your ideal customers.

Go-to-Market: A comprehensive strategic plan outlining how your product will reach your target customers. It includes sales channels, messaging, pricing, and tactics to achieve your commercial goals.

Product-Market Fit: The validation that your product satisfies a strong market demand and meets the needs of your target customers.

How do product experience, user experience, and customer experience differ?

These terms are often used interchangeably, but in this book they represent different expressions of the same idea. What matters most is creating integrated experiences rather than treating them as separate silos.

Product Experience: How customers experience your product across the entire journey, from discovery and onboarding to everyday use and ongoing support.

User Experience (UX): The design, usability, and functionality that bring your product experience to life in each customer interaction.

Customer Experience: The sum of all interactions customers have with your product and brand. In this framework, customer experience and product experience are used interchangeably to emphasize integrated thinking.

What's the difference between brand marketing, performance marketing, and growth marketing?

Brand marketing, performance marketing, and growth marketing represent complementary approaches that work together in an integrated marketing strategy. Each has a distinct focus but contributes to overall business growth.

Brand Marketing: Focuses on building awareness, perception, and loyalty across all touchpoints. Brand marketing encompasses both upper-funnel media channels (TV, online video, out of home) and any touchpoint that establishes and strengthens customer relationships—including product experience, website, app, packaging, PR events, and community engagement.

Performance Marketing: Focuses on generating immediate results through lower-funnel channels. Performance marketing is driven by specific metrics (clicks, conversions, sales) and typically employs channels like paid search, display ads, and direct response campaigns.

Growth Marketing: Blends elements of brand, product, and performance marketing but is distinctively characterized by a rapid test-and-learn approach. There's no universal definition of growth marketing, as organizations often define it based on their specific needs and objectives, with some companies using it to describe channel marketing and others to describe conversion optimization or any area that drives growth.

What's the role of product marketing? How does it differ from product management and other functions?

In early-stage companies or single-product businesses, roles like brand marketing, product marketing, and product management often blur. As organizations grow and product portfolios expand, these functions tend to specialize. However, without clear ownership and coordination, specialization can lead to unnecessary fragmentation and inefficiency.

Product Marketing: Serves as the bridge between product development and go-to-market activities. Product marketing translates product capabilities into customer-focused value propositions, develops positioning and messaging, enables sales and support teams, and orchestrates product launches.

Product Management: Focuses on building the right product and features. Product managers define the product vision, prioritize features, manage development, and work with design and engineering teams to bring products to life.

Acknowledgments

When I began the journey to write this book, I didn't know where it would lead me. I wrote the book that I wish had existed for me and my colleagues as we tried to launch transformative products and services. I'm beyond grateful to see it come to life.

The heart of this book is the love and generosity of my mentors, colleagues, friends, industry peers, editor, and publisher—all of whom were happy to share hard-won lessons to help people and companies succeed. In my marathon of over two years, it felt like this book would never be finished. But to me, the journey has been the biggest reward. I'm deeply grateful to so many who poured their heart and soul into sharing their wisdom, knowledge, and passion.

Thank you to Colin Raney for encouraging me to write in the first place. This book wouldn't exist without his encouragement to pursue my deepest desire to express myself and solve the hard challenges I've seen and experienced. Thanks for the inspiration and support.

Many of the colleagues and great friends from my time at Amazon read my early drafts, shared their knowledge, and helped me clarify my ideas. Thanks to Lauren Kim for being part of my entire journey—from the very first idea to the critical decisions along the way. Thanks to Simon Cassels, Marc Patijaud, Jeff Zaremski, Jenn Harvey, Mike Lemmon, Hannah Sturm, Stanley Chen, Jessica Bardoulas, Karina Kornacka, James Sheak, Diana Hua, Jacquelyn Miller, and many others for their contribution.

Special thanks to TJ Parker and Elliot Cohen, founders of PillPack, for their generous sharing of knowledge and insights into the startup world. Thanks to Dylan Field and Michael Amodeo from Figma; Eric Grunbaum from Airbnb; Megan Kang and Harry Porter from Rivian; and Rama Katkar, Shoshana Berger, Brooks Hocog, and Hilary Shirazi from Notion for providing support and essential details for the brand case studies. Thanks to Emmanuel Marot and Nancy King from the product and marketing community for sharing their industry knowledge and expertise.

Thanks to Marc-Antoine Jarry, Ben Butler, Mollie Partesotti, and Ravi Khanna, whom I worked with at Apple's Media Arts Lab. We had many conversations about industry challenges and opportunities, and I'm grateful for all the encouragement and great ideas. Professor Ben Lee, Professor Dennis Schorr, adviser Susan Zhang, and graduate students from the University of Southern California also provided valuable inputs and feedback.

My dear friends Sophie Ho, Peggie Li, Kristin Maile, YangYang Cheng, Jay Koottarappallil, and Jing Zhang read my early drafts and gave me honest feedback, pushing my thinking further and encouraging me to simplify and sharpen the book's focus. Mu Hu, Michael Cai, Mary Catherine Fixel, Yisha Zhang, Eva Chapiteau, and Robin Akashi supported me in their own ways. Thanks to them all.

There are special people you meet in life who change the trajectory of your career. For me, that's Mark Miller, who introduced me to Apple's

Media Arts Lab and later opened doors for me in the publishing world. Thank you, Mark.

I'm deeply grateful to Will Weisser for helping shape my book proposal, guiding me through the traditional publishing process, and making my early ideas sharper and my full manuscript shine. Will's thoughtful guidance helped me discover and establish my authentic voice.

Will also introduced me to Matt Holt, the publisher of his own imprint at BenBella Books. Thanks to Matt for having faith in me as a first-time author. Special thanks to my editor, Lydia Choi, whose editorial expertise and guidance helped shape every page of this book. I was so fortunate to collaborate with the BenBella team, including Katie Dickman, Brigid Pearson, Morgan Carr, Jessika Rieck, Mallory Hyde, Kerri Stebbins, Kathleen Berry-Li, Ariel Jewett, Susan Welte, and Kim Broderick. Thanks to them all for their dedication and expertise to bring this book to life. The level of support BenBella gives its authors is impeccable.

I feel lucky to have met Matteo Vianello through this book. He not only designed a beautiful, memorable cover but also provided incredible support that shaped my journey in meaningful ways. Matteo has been a true champion of the book, and I deeply appreciate his generosity. I also want to thank his team at Squero—Eleonora Fantin, Dan Stroud, and Philippa Adams—who elevated the book with exceptional design expertise.

On a personal note, none of this would have been possible without my family. Thanks to Hongzheng and Luke for giving me space and time so I could focus on writing, and for their encouragement throughout this journey. I'm also deeply grateful to my parents, who taught me discipline, grit, and hard work. Their selfless love has always inspired me to keep going and never stop dreaming.

There are so many people who contributed to the book, and I

apologize if I have missed any names. This has been a wonderful journey—something I never thought would be possible. Thank you all for making it possible.

Onward,

Lifang

PS: This book is just the beginning of the conversation. If you have feedback, questions, or ideas, I'd love to hear from you. Email me at hello@brandpowerbuiltin.com.

Notes

Introduction

1 "Ellen Couldn't Believe This Amazon Service Is Real," YouTube video,
   00:59, posted by TheEllenShow, August 22, 2018, https://www.youtube.com/
   watch?v=07kkUVphTFg&t=15s.
2 Kaushik Mani, LinkedIn post, March 2025, https://www.linkedin.com/posts/
   kaushik-mani-970b57b_retcon-activity-7306006825512902656-qyfd/.
3 Clayton M. Christensen, Scott Cook, and Taddy Hall, "Marketing Malpractice:
   The Cause and the Cure," *Harvard Business Review*, December 2005, https://hbr
   .org/2005/12/marketing-malpractice-the-cause-and-the-cure.
4 "Steve Jobs: Technology & Liberal Arts," YouTube video, 00:17, posted by Dana
   Peters, October 6, 2011, https://www.youtube.com/watch?v=KlI1MR-qNt8.

Chapter 1

1 "The Sharks Struggle to Understand Doorbot's Value," YouTube video, 00:58,
   posted by Shark Tank Global, April 15, 2022, https://www.youtube.com/
   watch?v=um-iVXiXedc&t=228s.
2 Simon Cassels, "Show Don't Tell, How Creativity Can Help See the Future,"
   Simon Cassels (blog), https://www.simoncassels.me/blog/show-dont-tell.
3 Ibid.
4 Jay Moye, "A Brief History of the Ring Video Doorbell and Its Evolution Over
   the Last 10 Years," About Amazon, May 2, 2023, https://www.aboutamazon.com/
   news/devices/a-brief-history-of-the-ring-video-doorbell-and-its-evolution-over
   -the-last-10-years.
5 Jay Moye, "A Brief History of the Ring Video Doorbell and Its Evolution Over
   the Last 10 years," About Amazon, May 2, 2023, https://www.aboutamazon.com/
   news/devices/a-brief-history-of-the-ring-video-doorbell-and-its-evolution-over
   -the-last-10-years.

6   Simon Cassels, "Show Don't Tell, How Creativity Can Help See the Future." Simon Cassels (blog), https://www.simoncassels.me/blog/show-dont-tell.

7   "Amazon and Ring Close Acquisition—Now Working Together to Empower Neighbors with Affordable Ways to Monitor Their Homes and Reduce Crime in Neighborhoods," Amazon Press Center, April 11, 2018, https://press .aboutamazon.com/2018/4/amazon-and-ring-close-acquisition-now-working -together-to-empower-neighbors-with-affordable-ways-to-monitor-their-homes -and-reduce-crime-in-neighborhoods.

8   Simon Cassels, "What People Are Saying," 2024, https://www.simoncassels.me/ awards.

9   Rob Gabriele, "2025 Home Security Market Report," SafeHome, February 11, 2025, https://www.safehome.org/resources/home-security-industry-annual/; "Ring Brand Awareness, Usage, Popularity, Loyalty, and Buzz Among Smart Home Users in the United States in 2023," Statista, March 12, 2024, https://www .statista.com/forecasts/1335446/ring-smart-home-brand-profile-in-the-united -states.

10   "This Is the Secret of Great Branding, According to Jeff Bezos," YouTube video, 0:13, posted by Michael Simmons, December 30, 2020, https://www.youtube .com/watch?v=8Jltq8vAxRI.

11   "Steve Jobs, Founder of Apple Speaks About Branding || Finance Advisor," YouTube video, 0:01, posted by Finance Advisor, February 3, 2024, https://www .youtube.com/watch?v=GtK98mAVTfo.

12   Geoffrey A. Moore, *Crossing the Chasm, 3rd Edition: Marketing and Selling Disruptive Products to Mainstream Customers* (New York: Harper Business, 2014).

13   "Advertising–United States," Statista Market Insights, October 2024, https:// www.statista.com/outlook/amo/advertising/united-states#ad-spending.

14   "Ad Spend Growth Tracks Ahead of the Economy: Dentsu Revise Up Global Ad Spend Growth Forecasts to 5.0% for 2024," Dentsu, May 29, 2024, https://www .dentsu.com/news-releases/ad-spend-growth-tracks-ahead-of-the-economy.

15   Arvind Hickman, "'Half of Advertising Is Dull, Ineffective and a Waste of Money'—System1's Jon Evans' Urgent Plea to Improve Creativity and Effectiveness," B&T, March 12, 2024, https://www.bandt.com.au/half-of -advertising-is-dull-ineffective-and-a-waste-of-money-system1s-jon-evans -urgent-plea-to-improve-creativity-and-effectiveness/.

16   James Wohr, "Ad Blocking: What It Is and Why It Matters to Marketers and Advertisers," Emarketer, September 4, 2024, https://www.emarketer.com/ learningcenter/guides/ad-blocking/.

17   "Google, Meta, and Amazon Grab 60% of $680bn Spend," DecisionMarketing, February 29, 2024, https://www.decisionmarketing.co.uk/news/google-meta-and -amazon-grab-60-of-680bn-spend.

18   "Snapchat CMO on Fixing Social Media, Being Creative Leader of the Decade and Lessons from Dan Wieden," YouTube video, 28:43, posted by Uncensored CMO, May 8, 2024, https://www.youtube.com/watch?v=QzN2Dl4mR0M.

19   Martin Guerrieria, "Revealed: The World's Most Valuable Brands of 2024," Kantar, June 12, 2024, https://www.kantar.com/inspiration/brands/revealed-the-worlds-most-valuable-brands-of-2024.

20   *2024 Most Valuable Global Brands*, Kantar, https://indd.adobe.com/view/publication/c21c3b44-92cb-4386-848e-c1641347979b/3arw/publication-web-resources/pdf/Kantar_BrandZ_2024_Most_Valuable_Global_Brands.pdf.

Chapter 2

1   Sir John Hegarty, LinkedIn post, August 2024, https://www.linkedin.com/posts/sir-john-hegarty-a1310a92_marketing-product-entrepreneurship-activity-7212406341779292162-E6lv/?utm_source=share&utm_medium=member_desktop.

2   "Experience Is Everything. Get It Right," PwC, https://www.pwc.com/us/en/services/consulting/library/consumer-intelligence-series/future-of-customer-experience.html.

3   "Five Questions with Neil Lindsay Vice-President Prime & Marketing, Amazon," Interbrand, https://interbrand.com/thinking/five-questions-with-neil-lindsay-vice-president-prime-marketing-amazon/.

4   Tony Fadell, *Build: An Unorthodox Guide to Making Things Worth Making* (New York: Harper Business, 2022).

5   Gitte Lindgaard, Gary Fernandes, Cathy Dudek, and Judith Brown, "Attention Web Designers: You Have 50 Milliseconds to Make a Good First Impression!" *Behaviour & Information Technology* 25(2), 115–126, ResearchGate, 2006, https://www.researchgate.net/publication/220208334_Attention_web_designers_You_have_50_milliseconds_to_make_a_good_first_impression_Behaviour_and_Information_Technology_252_115-126/citation/download.

6   Dylan Field, "Meet Us in the Browser," Figma, December 9, 2020, https://www.figma.com/blog/meet-us-in-the-browser/.

Chapter 3

1   Jon Linkov, "Most and Least Loved Car Brands," Consumer Reports, December 5, 2024, https://www.consumerreports.org/cars/car-reliability-owner-satisfaction/most-and-least-liked-car-brands-a1291429338/.

2   RJ Scaringe, interviewed by Rich Roll, "Building the Future," *The Rich Roll Podcast*, https://www.richroll.com/podcast/rj-scaringe-890/.

3   Kirsten Korosec, "Rivian Debuts an Electric Pickup and SUV Designed to Look Good While Getting Dirty," Tech Crunch, November 27, 2018, https://techcrunch.com/2018/11/27/rivian-debuts-an-electric-pickup-and-suv-designed-to-look-good-while-getting-dirty/.

4   "Rivian R1T Earns Highest Satisfaction Ranking of Any Vehicle in 2023 J.D. Power U.S. Electric Vehicle Experience Ownership Study," Rivian, February 28, 2023, https://rivian.com/newsroom/article/rivian-r1t-earns-highest-satisfaction-ranking-by-jd-power.

5   David A. Aaker, *Building Strong Brands* (New York: Free Press, 1995).

6   James Ryseff, Brandon F. De Bruhl, Sydne J. Newberry, "The Root Causes of Failure for Artificial Intelligence Projects and How They Can Succeed," RAND, August 13, 2024, https://www.rand.org/pubs/research_reports/RRA2680-1.html.

7   David Cahn, "AI's $600B Question," Sequoia Capital, June 20, 2024, https://www.sequoiacap.com/article/ais-600b-question/.

8   "Steve Jobs – Start with the Customer Experience," YouTube video, 00:25, posted by Paolo Landoni ENG, February 12, 2022, https://www.youtube.com/watch?v=QGIUa2sSYFI.

9   Matthew Woodward, "Airbnb Statistics [2025]: User & Market Growth Data," Search Logistics, March 17, 2025, https://www.searchlogistics.com/learn/statistics/airbnb-statistics/.

10 "Bluesky had 30 million users in Jan 2025," Bluesky post, January 28, 2025, 8:21PM, https://bsky.app/profile/did:plc:z72i7hdynmk6r22z27h6tvur/post/3lgu4lg6j2k2v?ref_src=embed&ref_url=https%253A%252F%252Fwww.theverge.com%252Fnews%252F602049%252Fbluesky-now-has-30-million-users.

11 Danny Sheridan, "Working Backwards at Amazon," Fact of the Day 1, Medium, January 19, 2020, https://medium.com/fact-of-the-day-1/working-backwards-at-amazon-a303c3680aa3.

Chapter 4

1   Dr. Marcus Collins, "Liquid Death's Billion-Dollar Valuation Stresses the Power of Brand," *Forbes*, March 20, 2024, https://www.forbes.com/sites/marcuscollins/2024/03/20/liquid-deaths-billion-dollar-valuation-underscores-the-power-of-brand/.

2   Evian Company Page, accessed September 22, 2025, https://www.evian.com/en_us/natural-spring-water/.

3   Fiji Company Page, accessed September 22, 2025, https://www.fijiwater.com/the-water.

4   Tom Huddleston Jr. and Zachary Green, "How Liquid Death's 40-Year-Old Founder Turned 'the Dumbest Name' and a Facebook Post into a $700 Million Water Brand," CNBC, November 26, 2022, https://www.cnbc.com/2022/11/26/liquid-death-ceo-mike-cessario-we-chose-the-dumbest-possible-name-for-water.html.

5   Tom Huddleston Jr., "42-Year-Old Turned a Facebook Page into a Real Company—Now It's Worth $1.4B and Spending Millions on a Super Bowl Commercial," CNBC, February 7, 2025, https://www.cnbc.com/2025/02/07/liquid-death-ceo-why-super-bowl-commercial-is-worth-the-money.html#:~:text=That%20approach%20helped%20Liquid%20Death,according%20to%20a%20company%20spokesperson.

6   Tom Huddleston Jr. and Zachary Green, "How Liquid Death's 40-Year-Old Founder Turned 'the Dumbest Name' and a Facebook Post into a $700 Million Water Brand," CNBC, November 26, 2022, https://www.cnbc.com/2022/11/26/

liquid-death-ceo-mike-cessario-we-chose-the-dumbest-possible-name-for-water.
html.

7   Roberto Verganti, *Design-Driven Innovation: Changing the Rules of Competition by
     Radically Innovating What Things Mean* (Boston: Harvard Business Press, 2009).

8   Simon Cassels, "Show Don't Tell, How Creativity Can Help See the Future,"
     Simon Cassels (blog), https://www.simoncassels.me/blog/show-dont-tell.

9   Daniel Kahneman, *Thinking, Fast and Slow* (New York: Farrar, Straus and Giroux,
     2013).

10   George Lakoff, *Metaphors We Live By* (Chicago: University of Chicago Press,
     2003).

11   Rae Yu, "The Evolution of the Xiaomi Brand: How It Conquered the Global
     Market with High Cost-Performance!" Tenten, November 15, 2023, https://
     tenten.co/learning/xiaomi-brand/.

12   "Mobile Vendor Market Share Worldwide," Statcounter GlobalStats, https://
     gs.statcounter.com/vendor-market-share/mobile.

13   Quibi's Open Letter, Medium, October 21 2020, https://quibi-hq.medium.com/
     an-open-letter-from-quibi-8af6b415377f.

14   Peter Thiel and Blake Masters, *Zero to One: Notes on Startups, or How to Build the
     Future* (New York: Crown Currency, 2014).

15   Avid Larizadeh, "The Ten Principles of Building Great Products," *Forbes*, May 23,
     2014, https://www.forbes.com/sites/avidlarizadeh/2014/05/23/ten-principles-on
     -the-journey-to-building-great-products/?utm_campaign=forbestwittersf&utm_
     source=twitter&utm_medium=social.

16   Alina Wheeler and Rob Meyerson, *Designing Brand Identity: A Comprehensive
     Guide to the World of Brands and Branding* (Hoboken, NJ: Wiley, 2024).

17   David A. Aaker, *Building Strong Brands* (New York: Free Press, 1995).

18   Brian Chesky, "Belong Anywhere," Medium, July 16, 2014, https://medium
     .com/@bchesky/belong-anywhere-ccf42702d010.

19   Brian Chesky, "Belong Anywhere," Medium, July 16, 2014, https://medium
     .com/@bchesky/belong-anywhere-ccf42702d010.

20   "Airbnb Live There," YouTube video, posted by Dose of Good Ads, May 19, 2022,
     https://www.youtube.com/watch?v=ddRBr2It00k.

Chapter 5

1   Kara Baskin, "The Entrepreneur's Journey: Elliot Cohen," Martin Trust Center,
     https://entrepreneurship.mit.edu/entrepreneurs-journey-elliot-cohen/.

2   Colin Raney, X post, November 29, 2022, 10:37 AM, https://x.com/colinraney/
     status/1597660958502985728.

3   "This Startup Revolutionized an Industry Through Design," IDEO, https://www
     .ideo.com/works/this-startup-revolutionized-an-industry-through-design.

4   "PillPack, 2013," Cooper Hewitt, https://collection.cooperhewitt.org/
     objects/1158817837/.

5   "TJ Parker: Building PillPack, the First E-Commerce Pharmacy, to Amazon's $1B

Acquisition," YouTube video, June 5, 2023, posted by 20VC with Harry Stebbings, https://www.youtube.com/watch?v=QgZDDGsZpx0.

6   Cees W. De Jong, ed., *Dieter Rams: Ten Principles for Good Design* (Munich: Prestel, 2021).

7   RJ Scaringe, interviewed by Rich Roll, "Building the Future," *The Rich Roll Podcast*, https://www.richroll.com/podcast/rj-scaringe-890/.

8   Dave Smith, "Apple Spent Hundreds of Hours Filming Flowers Blooming Just to Put on the Watch," *Business Insider*, April 8, 2015, https://www.businessinsider.com/apple-watch-faces-have-insane-level-of-detail-2015-4.

9   Amazon Pharmacy, How It Works, Amazon, https://pharmacy.amazon.com/how-it-works.

10   Carly Ayres, "Navigating the Intersection of Design and Business: A Conversation with Airbnb's Brian Chesky," Figma, August 10, 2023, https://www.figma.com/blog/config-brian-chesky-airbnb/.

11   Lenny Rachitsky, "Brian Chesky's New Playbook," Lenny's Newsletter, November 12, 2023, https://www.lennysnewsletter.com/p/brian-cheskys-contrarian-approach.

Chapter 6

1   Ivan Zhao, "100 Million of You," Notion HQ, September 3, 2024, https://www.notion.com/blog/100-million-of-you.

2   Kenrick Cai, "$10 Billion Productivity Startup Notion Wants to Build Your AI Everything App," *Forbes*, April 11, 2024, https://www.forbes.com/sites/kenrickcai/2024/04/11/10-billion-productivity-startup-notion-wants-to-build-your-ai-everything-app/.

3   Carmel DeAmicis, "Design on a Deadline: How Notion Pulled Itself Back from the Brink of Failure," Figma, March 6, 2019, https://www.figma.com/blog/design-on-a-deadline-how-notion-pulled-itself-back-from-the-brink-of-failure/.

4   David Pierce, "The Only App You Need for Work-Life Productivity," *Wall Street Journal*, March 21, 2018, https://www.wsj.com/articles/the-only-app-you-need-for-work-life-productivity-1521640800?gaa_at=eafs&gaa_n=ASWzDAhj_G5qfy4wB-S2n1PPYEG3weg0JQgriw4fu3DU8cC-2Ef5ImjEVWaNOzI4XbY%3D&gaa_ts=68c06f24&gaa_sig=pXQ9Enu8LIyNkJgUj2hX8HayXZ8piLzh-9pRs8iyWlIAFmbz_xnG.

5   Kenrick Cai, "$10 Billion Productivity Startup Notion Wants to Build Your AI Everything App," *Forbes*, April 11, 2024, https://www.forbes.com/sites/kenrickcai/2024/04/11/10-billion-productivity-startup-notion-wants-to-build-your-ai-everything-app/.

6   Lucy Woolfenden, "Notion's Rise to 90% Organic Growth," LinkedIn post, November 21, 2024, https://www.linkedin.com/pulse/notions-rise-90-organic-growth-lucy-woolfenden-cgvqe/.

7   Jason Feifer, "Meet the Millennial Founder Who Built a $10 Billion Startup on an Ancient Philosophy: 'There's No Better System Than Nature,'" *Entrepreneur*, January 17, 2025, https://www.entrepreneur.com/leadership/the-ancient -philosophy-behind-a-10-billion-startup/481661.

8   Megan Finnerty, "Augmenting Human Intellect, No Code Required," Sequoia Capital, October 13, 2022, https://www.sequoiacap.com/article/notion-spotlight/.

9   "Introducing Apple Intelligence, the Personal Intelligence System That Puts Powerful Generative Models at the Core of iPhone, iPad, and Mac," Apple Press Release, June 10, 2024, https://www.apple.com/newsroom/2024/06/introducing -apple-intelligence-for-iphone-ipad-and-mac/.

10   Kyle Wiggers, "Apple Faces Lawsuit over Apple Intelligence Delays," TechCrunch, March 20, 2025, https://techcrunch.com/2025/03/20/ apple-faces-lawsuit-over-apple-intelligence-delays/?guccounter=1&guce_ referrer=aHR0cHM6Ly93d3cuZ29vZ2xlLmNvbS8&guce_referrer_sig=AQAA AD5ZrHzl27MIIgvPqbBKhEHH2VGzYmyQD8qMFN0EwSvttCtUp8vi4P vc-_S_Miv-WVcq3Zee2tC7Wbmem1Fm5GixmX4DE-LWJsI6CBRlkWJRo -L6GVf0268j0_ShHDb2GoPDuexhZVT0V1a7qODdKcrnNvY6P-_zA_ xrZrO1JTHd.

11   Sam Blum, "Bird Led the Charge into Micromobility. Now It's Seeking Bankruptcy Protection," *Inc.*, December 20, 2023, https://www.inc.com/sam -blum/bird-led-the-charge-into-micromobility-now-its-bankrupt.html.

12   "Bird Successfully Emerges from Bankruptcy as a Stronger Company and Will Operate as the Global Anchor Brand of Newly Established Third Lane Mobility Inc.," *Bird Cities Blog*, April 5, 2024, https://www.bird.co/blog/bird-successfully -emerges-from-bankruptcy-as-a-stronger-company-and-will-operate-as-the -global-anchor-brand-of-newly-established-third-lane-mobility-inc/.

13   "Waymo One is now open to all in Los Angeles," Waymo, November 12, 2024, https://waymo.com/blog/2024/11/waymo-one-open-to-all-in-los-angeles.

14   Matt Ridley, *How Innovation Works: And Why It Flourishes in Freedom* (New York: Harper, 2020).

15   "Notion's Lost Years, Near Collapse During COVID, Staying Small to Move Fast, Building Horizontal," YouTube video, posted by Lenny's Podcast, March 6, 2025, https://www.youtube.com/watch?v=IIPKMixTMfE.

16   "Steve Jobs Introducing the iPhone at MacWorld 2007," YouTube video, 2:29, December 2, 2010, posted by superapple4ever, https://www.youtube.com/ watch?v=x7qPAY9JqE4.

17   Roger L. Martin, Jann Schwarz, and Mimi Turner, "The Right Way to Build Your Brand," *Harvard Business Review*, January–February 2024, https://hbr.org/2024 /01/the-right-way-to-build-your-brand.

18   "Apple 'There's an App for That' Campaign," 2010 Gold Consumer Electronics, Effie, https://www.effie.org/cases/theres-an-app-for-that/.

Chapter 7

1   "Keeping Customers Informed Through Market Volatility," *Robinhood Blog*, January 28, 2021, https://newsroom.aboutrobinhood.com/keeping-customers -informed-through-market-volatility/.

2   Ron Shevlin, "The GameStop Aftermath: The Rise of the Anti-Robinhood," *Forbes*, February 8, 2021, https://www.forbes.com/sites/ronshevlin/2021/02/08/ the-rise-of-the-anti-robinhood/.

3   Nathaniel Popper & Matt Phillips, "In GameStop Saga, Robinhood Is Cast as Villain," *New York Times*, February 18, 2021, https://www.nytimes.com/2021 /02/18/business/gamestop-robinhood-hearing.html.

4   "Robinhood ft. Vlad Tenev – Reinventing Finance for a New Generation," Sequoia Crucible Moments: Episode 16, https://www.sequoiacap.com/podcast/ robinhood-ft-vlad-tenev-reinventing-finance-for-a-new-generation/.

5   Michael de la Merced and Erin Griffith, "Robinhood Is Fined $70 Million over Misleading Customers and System Outages," *New York Times*, June 30, 2021, https://www.nytimes.com/2021/06/30/technology/robinhood-fined-misleading -customers.html.

6   "United States District Court Northern District of California San Jose Division," Classaction, https://www.classaction.org/media/cezana-v-robinhood-financial-llc -et-al.pdf.

7   "Robinhood ft. Vlad Tenev – Reinventing Finance for a New Generation," Sequoia Crucible Moments: Episode 16, https://www.sequoiacap.com/podcast/ robinhood-ft-vlad-tenev-reinventing-finance-for-a-new-generation/.

8   Jeffrey F. Rayport, Davide Sola, and Martin Kupp, "The Overlooked Key to a Successful Scale-Up," *Harvard Business Review*, January–February 2023, https:// hbr.org/2023/01/the-overlooked-key-to-a-successful-scale-up.

9   "How Brand Building and Performance Marketing Can Work Together," *Harvard Business Review*, May 2023, https://hbr.org/2023/05/how-brand%20building -and-performance-marketing-can-work-together.

10   "A Message from Co-Founder and CEO Brian Chesky," Airbnb, May 5, 2020, https://news.airbnb.com/a-message-from-co-founder-and-ceo-brian-chesky/.

11   "Airbnb Q4 2020 Earnings Call," Call Street, February 25, 2021, https://s26. q4cdn.com/656283129/files/doc_financials/2020/q4/Airbnb,-Inc .Q42020EarningsCall-(1).pdf.

12   Gideon Spanier, "Airbnb Slashes Spend in Permanent Shift from Performance Marketing to Brand," Campaign Asia, March 2, 2021, https://www.campaignasia .com/article/airbnb-slashes-spend-in-permanent-shift-from-performance -marketing-to-brand/468053.

13   Simon Boddy, "Airbnb: A Crisis PR Success Story," Ambitious PR, May 20, 2022, https://www.ambitiouspr.co.uk/airbnb-pr-crisis/.

14   "Q4 2022 Shareholder Letter," The SEC, https://www.sec.gov/Archives/edgar/ data/1559720/000119312523039008/d451233dex991.htm.

15   "Marketing Expenses of Booking Holdings from 2015 to 2024," Statista, https://
     www.statista.com/statistics/225474/booking-holdings-operating-expenses/.

16   "Brand Marketing Pays Off for Airbnb," WARC, November 3, 2022, https://www
     .warc.com/content/feed/brand-marketing-pays-off-for-airbnb/en-GB/7437.

17   Megan Graham, "Airbnb Says Its Focus on Brand Marketing Instead of Search Is
     Working," *Wall Street Journal*, November 3, 2022, https://www.wsj.com/articles
     /airbnb-says-its-focus-on-brand-marketing-instead-of-search-is-working
     -11667506438.

18   Matt Alagiah, "How Spotify's Wrapped Campaign for 2022 Came Together," It's
     Nice That, November 30, 2022, https://www.itsnicethat.com/features/spotify
     -wrapped-campaign-identity-2022-graphic-design-301122.

19   Tejas Manohar, "The Science Behind Spotify Wrapped: Tracking 500M Users,"
     hightouch, November 30, 2023, https://hightouch.com/blog/how-spotify
     -wrapped-works.

20   Jeff Beer, "Airbnb's New Icons Feature Lets You Book a Stay in Pop Culture's
     Most Famous Locations," *Fast Company*, May 1, 2024, https://www.fastcompany
     .com/91115714/airbnbs-new-icons-feature-lets-you-book-a-stay-in-pop-cultures
     -most-famous-locations.

21   "Notion Ambassador Program," https://workshops.aucklandimprovmarathon.co
     .nz/notion/Notion-Ambassador-Program-40c3b3ee8c744e7fad34ab4ac9765773.

22   Lucy Woolfenden, "Notion's Rise to 90% Organic Growth," LinkedIn post,
     November 21, 2024, https://www.linkedin.com/pulse/notions-rise-90-organic
     -growth-lucy-woolfenden-cgvqe/.

23   Rachel Botsman, *Who Can You Trust?: How Technology Brought Us Together and
     Why It Might Drive Us Apart* (New York: PublicAffairs, 2017).

Chapter 8

1   Lily Meier, "Nike's Biggest Drop on Record Puts Pressure on CEO
     Donahoe," Bloomberg, June 28, 2024, https://www.bloomberg.com/news/
     articles/2024-06-28/nike-s-biggest-drop-in-23-years-raises-pressure-on-ceo
     -donahoe?embedded-checkout=true.

2   "Nike, Inc. Reports Fiscal 2024 Fourth Quarter and Full Year Results," Nike,
     June 27, 2024, https://investors.nike.com/investors/news-events-and-reports/
     investor-news/investor-news-details/2024/NIKE-Inc.-Reports-Fiscal-2024
     -Fourth-Quarter-and-Full-Year-Results/default.aspx.

3   Massimo Giunco, "Nike: An Epic Saga of Value Destruction," LinkedIn post,
     July 28, 2024, https://www.linkedin.com/pulse/nike-epic-saga-value-destruction
     -massimo-giunco-llplf/.

4   "Nike Announces Senior Leadership Changes to Unlock Future Growth Through
     the Consumer Direct Acceleration," Nike, July 22, 2020, https://investors
     .nike.com/investors/news-events-and-reports/investor-news/investor-news
     -details/2020/Nike-Announces-Senior-Leadership-Changes-to-Unlock-Future
     -Growth-Through-the-Consumer-Direct-Acceleration/default.aspx.

5   Ashley Rodriguez, "Nike Is Mending Relationships with Retailers as More Brands Recognize the Constraints of Direct-to-Consumer Models," *Business Insider*, June 11, 2023, https://www.businessinsider.com/nike-expands-deals-with-some-retailers-amid-dtc-shift-in-strategy-2023-6.

6   Massimo Giunco, "Nike: An Epic Saga of Value Destruction," LinkedIn post, July 28, 2024, https://www.linkedin.com/pulse/nike-epic-saga-value-destruction-massimo-giunco-llplf/.

7   David A. Aaker, *Building Strong Brands* (New York: Free Press, 1995).

8   Amazon 2009 Letter to Shareholders, https://s2.q4cdn.com/299287126/files/doc_financials/annual/AMZN_Shareholder-Letter-2009-(final).pdf.

9   Lenny Rachitsky, LinkedIn post, February 2025, https://www.linkedin.com/posts/lennyrachitsky_we-believe-that-having-metrics-make-the-activity-7293680963073495040-IPta/?utm_source=share&utm_medium=member_desktop.

Chapter 9

1   Dylan Field, "Meet Us in the Browser," Figma, December 9, 2020, https://www.figma.com/blog/meet-us-in-the-browser/.

2   United States Securities and Exchange Commission, July 21, 2025, https://www.sec.gov/Archives/edgar/data/1579878/000162828025035381/figma-sx1a.htm.

3   "Great Place to Work in 2021," September 2021, https://www.greatplacetowork.com/certified-company/7015385.

4   "Figma Uses iVerify to Protect People-First Design with People-First Security," iVerify, https://iverify.io/reports/figma.

5   Laura Pang, "How We Engineer Feedback at Figma with Eng Crits," Figma, February 29, 2024, https://www.figma.com/blog/how-we-run-eng-crits-at-figma/.

6   Nilay Patel, "Why Figma CEO Dylan Field Is Optimistic About AI and the Future of Design," *The Verge*, March 18, 2024, https://www.theverge.com/24102160/figma-ceo-adobe-deal-design-ai-web-future-regulation-sxsw-decoder-interview.

7   Daniel Coyle, *The Culture Code: The Secrets of Highly Successful Groups* (New York: Random House Business Books, 2019).

8   Daniel Coyle, "An Excerpt from The Culture Code," https://danielcoyle.com/excerpt-culture-code/.

9   Phil Knight, *Shoe Dog: A Memoir by the Creator of Nike* (New York: Scribner, 2016).

10   Jim Collins, *Good to Great: Why Some Companies Make the Leap . . . and Others Don't* (New York: Harper Business, 2001).

11   Tom Eisenmann, *Why Startups Fail: A New Roadmap for Entrepreneurial Success* (New York: Crown Currency, 2021).

12   Christina Farr, "At Amazon's PillPack Pharmacy, New Hires Do 'Empathy Training' by Sorting Pills While Wearing Bulky Gloves and Glasses," CNBC, September 21, 2019, https://www.cnbc.com/2019/09/21/amazon-pillpack-empathy-training-shows-employees-what-customers-feel.html.

13 Niki Jorgensen, "Beyond Compensation and Benefits: Why Company Culture Is Key," *Forbes*, April 14, 2022, https://www.forbes.com/councils/forbeshumanresourcescouncil/2022/04/14/beyond-compensation-and-benefits-why-company-culture-is-key/.

14 "Median Tenure with Current Employer Was 3.9 Years in January 2024," US Bureau of Labor, October 11, 2024, https://www.bls.gov/opub/ted/2024/median-tenure-with-current-employer-was-3-9-years-in-january-2024.htm#:~:text=The%20median%20number%20of%20years,not%20available%20for%20every%20year.

15 Daniel Coyle, *The Culture Code: The Secrets of Highly Successful Groups* (New York: Random House Business Books, 2019).

16 "Peter Drucker On Marketing," *Forbes*, July 3, 2006, https://www.forbes.com/2006/06/30/jack-trout-on-marketing-cx_jt_0703drucker.html.

17 Dara Khosrowshahi, "Uber's New Cultural Norms," LinkedIn post, November 7, 2017, https://www.linkedin.com/pulse/ubers-new-cultural-norms-dara-khosrowshahi/.

18 Megan Rose Dickey, "These Are Uber's New 'Cultural Norms,'" Tech Crunch, November 7, 2017, https://techcrunch.com/2017/11/07/these-are-ubers-new-cultural-norms/.

19 "AI's Impact on Marketers and Brands, According to 500 CMOs," Frontify, https://www.frontify.com/en/blog/ai-impact-on-marketers-and-brands.

20 Ori Faran, "What If Customers Don't Want Your AI Chatbot?" *Forbes*, August 1, 2024, https://www.forbes.com/councils/forbesbusinesscouncil/2024/08/01/what-if-customers-dont-want-your-ai-chatbot/.

21 Nilay Patel, "Why Figma CEO Dylan Field Is Optimistic About AI and the Future of Design," *The Verge*, March 18, 2024, https://www.theverge.com/24102160/figma-ceo-adobe-deal-design-ai-web-future-regulation-sxsw-decoder-interview.

Conclusion

1 Erin Griffith and Tripp Mickle, "HP to Buy Humane, Maker of the Ai Pin, for $116 Million," *New York Times*, February 18, 2025, https://www.nytimes.com/2025/02/18/technology/hp-humane-ai-pin.html.

2 Ai Pin, Kameron Burk's website, https://kameronburk.com/Ai-Pin-1.

Index

A

Aaker, David A., 54, 87, 173
accountability, 204–205
acquisition. *See* customer acquisition
acquisition economics, 133–134, 138–139
Adobe, 192, 194
advertising
    brand building through myth, 24–27
    creative work investment in, 108
    function of, 136–137
    relevance of, 137
    saturation of, 25
    statistics regarding, 25, 26
    as your product, 109–112, 115
    See also marketing
advocacy, amplifying organic growth through, 161–164, 167
Ai Pin (Humane), 3–4, 211–213
Airbnb
    approach of, 35
    assessment of, 177
    brand building focus of, 177
    brand identity of, 89, 154
    brand marketing of, 26, 154–155
    company strategy of, 113
    connections of, 39
    COVID-19 pandemic and, 152–155
    culture of, 199
    customer insights of, 62–63
    design team of, 113
    evolution of, 62–63
    fragmentation challenge of, 113
    icons of, 162

    passion and purpose of, 82
    performance marketing of, 153, 154–155
    positioning strategy of, 56, 62–63
    public relations and, 162
    rebranding of, 90–91
    statistics of, 62, 153
    traffic sources for, 177
Amazon
    Ads, 26
    brand building by, 37
    brand health metrics of, 183
    brand identity of, 86–87
    culture of, 195–196
    evolution of, 88
    integration process of, 206
    Key In-Home Delivery, 1–3, 21, 124–125
    marketing of, 110–112
    One Medical, 92
    Pharmacy, 85–86, 99–100, 110–112, 157
    PillPack, 99–100, 105, 106, 199, 201
    Prime, 85–86
    rebranding of, 92–93
    repeatable mechanisms of, 114
    working backwards mechanism of, 66, 196
    See also Ring
Apple
    advertising by, 137
    App Store, 136
    brand expression of, 156
    branding of, 5
    connections of, 39

culture of, 195
customer benefits of, 59
Intelligence, 125
iPhone, 4, 20, 38, 125, 135, 136, 137
Mac computer of, 23–24
marketing and advertising as product
    of, 109
product development process of, 59
storytelling by, 135, 136
user experience (UX) and, 38
Watch, 60, 108–109
artificial intelligence (AI)
    $600B Question of, 58–59
    challenges of, 22–23
    customer relationships and, 215
    in customer service, 208
    ethics in, 106
    human insights and, 67–68
    impacts of, 54–55
    leading through, without losing
        customer focus, 207–209, 210
    pitfalls of, 58
    quick information vs. strategic insights
        and, 67–68
    technology-first approach in, 58–59
attribution, problem of, 178–179
awareness, 110, 111, 124, 175, 176, 177, 182

B

benchmarking, competitive, 65–66, 74
Bezos, Jeff, 18, 66, 86–87, 181
Bhatt, Baiju, 145
Bird, 126
Bluesky, 63
Booking Holdings, 26, 154
Botsman, Rachel, 165
brand
    as bridge, 23–24
    building through advertising myth,
        24–27
    business performance metrics
        integration with, 181–184, 188
    connections through, 4
    defined, 17–19, 219–220
    foundations of, 79–80
    function of, 6, 36
    as greatest asset, 208–209
    as at heart of products, 213–215
    integration of, into product experience,
        45
    myths of undervaluing, 16–27, 28–29
    new approach for, 27–28, 29

performance marketing integration with,
    151–156, 167
product as inseparable with, 19–21, 29
product innovation as success driver over
    myth, 21–24
product integration with, 81
protection of, 207–208
as separate from product myth, 17–21
statistics regarding, 27
storytelling and, 22
trade-offs of, 31–33
as trust, 18
value, 37–38
brand architecture, 88–89, 94
brand building
    building blocks for, 49
    customer-centric approach in, 40
    defined, 18, 220
    foundation of, 155–156
    overview of, 20–21
    product experience in, 36–40, 47
    purpose of, 187
    selling points for, 156
    as shared responsibility, 7, 42–44, 47
    See also specific elements
brand development, positioning for, 71
brand equity, 173–176, 187–188, 220
brand health metrics, 175
brand identity
    balancing bold vision with reality in,
        83–84, 93
    brand architecture and, 88–89, 94
    bringing vision to life through story and,
        80–84, 93
    consumer behaviors and, 77–78
    defined, 18, 220
    elements of, 87
    establishing, 44–45
    foundations of, 79–80
    function of, 127
    as long-term asset, 86–88, 93
    naming decision in, 84–86, 93
    overview of, 75
    rebranding and, 89–93, 94
    repositioning and, 89–93, 94
    starting with why for, 81–82
brand marketing, 154–155, 157, 167, 223. See
    also marketing
brand positioning, 54–55, 56–57, 61–66,
    68–70, 74. See also positioning strategy
brand power, 46, 147–148, 166, 215–217, 219

Brand Power Built In approach
    building blocks of, 45, 47, 50, 118
    defined, 219
    implementation of, 44–46
    integration in, 40
    overview of, 34–36
    principles of, 35, 47
    See also specific elements
brand thinking, 34, 39–40, 41, 100–102
branding, defined, 18, 220
build-measure-learn feedback loop, 184
business performance metrics, 174–175,
    181–184, 188
business strategy, positioning for, 70

C

campaign approach, 18, 155–156, 167,
    176–179, 187–188, 220
category convention, 54, 73
Cessario, Mike, 76–77
change, sustaining culture through, 205–207,
    210
channel activation, 156
Chesky, Brian, 91, 113, 154
China, EV competition in, 55
Christiansen, Richard, 158
Coca-Cola, 20, 24
Cohen, Elliot, 22, 96
collaboration, 69, 119–122, 192–195
Collins, Jim, 198
communication, 19, 107–108, 158–160
community, 121–122, 161–164, 167
competitive benchmarking, 65–66, 74
connections, 4, 82, 106–109, 115, 160–161,
    212–213
Consumer Direct Acceleration strategy,
    170–171
COVID-19 pandemic, 152–155
Coyle, Daniel, 195, 197
creative execution, 156
cross-functional
    alignment in teams, 177, 183, 204
    brand building as, 7
    collaboration, 43, 68, 200, 202
    communication, 159
    marketing, 110
Cuban, Mark, 14
culture
    accountability for, 204–205
    anchoring around North Star, 196–198,
        209
    brand vision in, 191

    clear roles and responsibilities in,
        202–203
    customer experience and, 195, 199
    customer-centric, 191
    defined, 195
    designing organizations to prevent
        fragmentation for, 201–205, 209
    Figma's story of, 192–195
    introduction to, 189, 191
    investment in humans in, 198–200, 209
    leading through AI without losing
        customer focus for, 207–209, 210
    maintaining growth and change in,
        206–207
    marketing at core of innovation and,
        200–201, 209
    mission and, 197
    people-centric, 193, 198–200, 209
    purpose and, 197
    as shaping products, 195–196, 209
    statistics regarding, 195, 199–200
    sustaining, 205–207, 210
    as wrong, 205–206
customer acquisition
    brand power and, 147–148, 166
    defined, 220
    integrating brand and performance
        marketing for, 151–156, 167
    leveraging, 149
    metrics of, 175
    sequencing strategy for, 131–134, 141
    shift from, 146–147
    through storytelling, 134–138, 141
customer experience (CX)
    cohesion to fragmentation in, 41
    communication and, 159–160
    connecting the dots for, 112–114, 115
    culture and, 195, 199
    defined, 222
    end-to-end walk-through of, 114
    fragmenting, 112–113
    investment in, 37
    misalignment of, 213
    starting with, 59
    See also product experience; user
        experience (UX)
customer relationship management (CRM),
    158–161
customer retention
    brand power and, 147–148, 166
    deepening customer connections
        through data for, 160–161

defined, 220
   focus on, 146–147
   leveraging, 149
   metrics for, 175
   strengthening customer relationships
      and, 158–161, 167
customer service, 72, 106, 122, 208
customers
   connection of, 212–213
   education of, 136–138
   email communication with, 41
   engagement by, 146–147, 149, 220
   evolving with, 186–187, 188
   expectations of, 106
   feedback of, 66, 130–131
   getting attention of, 132–133
   insights of, 61–66
   journey of, 40–41, 47, 110, 111
   lifecycle of, 220
   loyalty of, 221
   needs of, 58–59
   niche segment of, 60
   obsession of, 106
   personas and profiles of, 65, 74
   pulling of, 124–125, 140
   relationships of, 37, 215
   relevance with, 186–187, 188
   segmentation, 64–65, 74, 128, 221
   trust of, 212–213
   uniting all functions around, 43–44
   value of, balancing efficiency with,
      207–208

D
data, 160–161, 169–173, 187
day one
   brand strategy and power from, 6, 11,
      13-16, 27, 28, 31, 33, 77, 219
   marketing from, 200
   product story on, 22
   pushing a product from, 127
   why from, 82
   winning from, 31, 33, 59
Decourcy, Colleen, 26
DeGeneres, Ellen, 2
differentiation, 147, 174
digital platforms/products, 25–27, 29, 39,
   43–44, 106, 208
Drucker, Peter, 180, 201
Dye, Alan, 108–109

E
Eisenmann, Tom, 58, 198–199
emotional connections, 106–109, 115
Evian, 76
Expedia Group, 26, 154
experience design, brand thinking to, 39–40
experimentation, embracing iteration and,
   184–186, 188
exploitation, exploration, and extrapolation,
   150

F
Facebook, 63, 85
Fadell, Tony, 37
feedback, customer, 66, 130–131
Field, Dylan, 43–44, 192, 193, 194, 208
Figma, 35, 43–44, 83, 192–195, 197
Flamingo Estate, 158–159
fragmentation, 112–113, 172, 201–205, 209
full-funnel campaign approach, 155–156,
   176–179, 187–188

G
GameStop, 144–145
Giunco, Massimo, 170
go-to-market (GTM) strategy, 71–72, 122,
   125, 131, 139, 140, 221–222. *See also*
   launching
growth
   amplifying through community and
      advocacy, 161–164, 167
   brand equity as engine for, 173–174, 187
   driving, 148–151, 167
   maintaining, 206–207
   organic, 132–133, 161–164, 166, 167
   paid, 133
   requirements for, 172, 173
   sequenced, 122–127
   stages of venture, 150
   sustainable, 46, 172, 173–174, 187
   sustaining culture through, 205–207,
      210
   trust as multiplier of, 165–166, 167
growth marketing, defined, 223

H
HBO, 89–90
Hegarty, Sir John, 36
Hodgman, John, 24
human psychology, branding and, 77–78
Humane, 83–84, 211–213

I

impact, measuring, 46
innovations
 adoption cycle of, 23
 in brand equity, 174
 bridging the gap of, 29
 challenges of, 21
 function of, 128
 marketing at core of, 200–201, 209
 prioritization for, 136
 storytelling for, 136
input-output framework, 180–181, 188
inseparable
 marketing and innovation as, 201, 209
 product and brand as, 6, 9, 19–21, 29
insights, 61–66, 67–68, 73
iPhone (Apple), 4, 20, 38, 125, 135, 136, 137
iteration
 brand equity in, 173–176
 connecting the dots through input-
  output framework for, 180–181, 188
 embracing experimentation and,
  184–186, 188
 evolving with customers to stay relevant
  for, 186–187, 188
 integrating brand and business
  performance metrics for, 181–184,
  188
 introduction to, 169
 measuring integrated, full-funnel
  campaigns for, 176–179, 187–188
 Nike's story of, 169–173

J

Jobs, Steve, 5, 18, 59, 109, 114, 135
Jun Lei, 82

K

Kalanick, Travis, 205
Khosrowshahi, Dara, 205–206
Knight, Phil, 198

L

Lakoff, George, 80
Last, Simon, 120
launching
 acquisition economics for, 138–139
 business priority in, 138
 customer feedback and, 130–131
 customer understanding in, 215
 expansion powered by community in,
  121–122

fast expansion dangers in, 126–127
finding product-market fit in, 128–131,
 141
lack of marketing in, 124
launch stage of, 139, 140
laying pre-launch foundation for,
 127–128, 141
misalignment with target audience in,
 124
Notion's story of, 119–122
overly broad targeting in, 124
overview of, 119
path to product-market fit in, 123
pitfalls of, 123–124
pivot in, 120–121
post-launch stage of, 139, 140
pre-launch stage of, 139
premature scaling in, 124
prioritizing owned and early channels
 for, 132–133
product readiness in, 138
pulling customers into, 124–125, 140
scale stage of, 139, 140
with sequenced go-to-market strategy,
 46
sequenced growth approach in, 122–127,
 138–140, 141
sequencing customer acquisition
 strategy for, 131–134, 141
stages of, 138–140, 141
storytelling for, 134–138, 141
target audience for, 138
as too big, 213
Lime, 126
Limp, Dave, 15
Lindsay, Neil, 37
Liquid Death, 75–79, 162–163
Long, Justin, 24
lower-funnel marketing, 177, 185, 187–188,
 221
Lütke, Tobias, 183

M

marketing
 at core of innovation, 200–201, 209
 cost and efficiency measuring for, 177
 experiential, 162
 function of, 34
 funnel, 221
 incremental testing in, 185
 lack of, in launching, 124
 measuring, 176–179, 187–188

positioning strategy and, 69
product as, 109–112, 115, 164
product integration with, 98–99
solving product design challenges
    through, 110–112
upper-funnel, 177, 185, 187–188, 221
See also advertising
messaging
brand development and, 57, 203
cohesion with, 107
design choice in, 77
GTM approach and, 128
inconsistent, 101
investment in, 107–108
positioning for, 56, 155–156
simplifying, 88, 94
storytelling in, 141
trust and, 166
Meta, 26, 85
mindset shift, 7, 43, 193
mission statement, function of, 69–70
Moore, Geoffrey A., 23, 54

N

naming, as one-way door decision, 84–86, 93
Nest Thermostat, 37
Net Promoter Score (NPS), 175, 221
Netflix, 61
Nike, 169–173, 184, 186–187, 198
North Star, 56–57, 69, 70, 73, 79, 196–198,
    209
Notion, 35, 119–122, 128, 130–131, 132,
    163–164, 166, 197–198

O

Ogilvy, David, 84
One Medical, 92
organic growth, 161–164, 166, 167. *See also*
    growth
organizational fragmentation, 201–205, 209
overpromising, 105–106

P

Parker, TJ, 96, 100
PepsiCo, 24
performance marketing, 151–156, 157, 167,
    223. *See also* marketing
PillPack, 35, 96–100, 105, 106, 199, 201
positioning strategy
alignment in, 69
areas of, 57
for brand development, 71

for business, 70
changes to, 61
competitive benchmarking and, 65–66,
    74
components of, 54, 73
customer personas and profiles of, 65, 74
customer segmentation and, 64–65, 74
for customer service, 72
defined, 73, 221
developing, 44
failing to stand out in the sameness and,
    60–61
focusing on technology over customer
    needs and, 58–59
40 questions to ask regarding, 70–74
function of, 127
for go-to-market, 71–72
importance of, 54–55
insight importance for, 67–68, 73
for launching, 127
marketing collaboration for, 69
mass appeal without purposeful focus
    and, 59–60
mission statement and, 69–70
as North Star, 56–57, 69, 70, 73
operationalizing across your
    organization, 68–70, 74
overview of, 51
pitfalls of, 58–61, 73
for product development, 70–71
Rivian story of, 51–53
story and, 79–80
turning insights into strategic
    opportunities and, 61–66
Powell, Matt, 172
product
advertising as, 109–112, 115
appeal of, 78
brand as at heart of, 213–215
brand as inseparable with, 19–21, 29
brand as separate from myth, 17–21
brand integration with, 81
brand thinking to, 34
broken trust with, 105–106
culture as shaping, 195–196, 209
as customer journey, 7, 47
delight in, 108–109
design, 100–104, 108, 110–112, 115,
    192–195, 208
development, 70–71
differentiation of, 147
as driving brand value, 37–38

product (*continuted*)
    education of, 136–138
    as entire customer journey, 40–41
    functionality of, 104
    innovation as success driver over brand
        myth, 21–24
    marketing as, 109–112, 115, 164
    marketing integration with, 98–99
    as meaningful, 21–23
    naming decision for, 80, 84–86
    niche for, 128–129, 141
    readiness of, 138
    showcasing, 136–137
    storytelling and emotional connections
        within, 106–109, 115
    uniting functions of, around customer,
        43–44
    value of, 85, 137
product experience
    avoiding overpromising and
        underdelivering in, 105–106
    brand equity and, 174
    brand thinking and, 100–102
    connecting the dots for, 112–114, 115
    defined, 222
    defining shared vision and design
        principles in, 102–104, 115
    delivering on your promise in, 104–106,
      115
    integration of brand into, 45
    as most powerful brand builder, 36–40,
      47
    PillPack story of, 96–100
    product marketing and, 25
    See also customer experience (CX); user
        experience (UX)
Product Hunt, 121
product management, defined, 224
product marketing, 25, 223–224. *See also*
    marketing
product-market fit, 55, 123, 128–131, 141,
    222
profit-market fit, 150
promise, 54, 73, 104–106, 115, 135–136, 141
public relations, 162–163

Q

Quibi, 83–84

R

Rams, Dieter, 101–102
Raney, Colin, 98–99

reasons to believe, 54, 73
rebranding, in brand identity, 89–93, 94
Reed, Andrew, 147
repositioning, in brand identity, 89–93, 94
Responsible, Accountable, Consulted, and
    Informed (RACI), 204
retention. *See* customer retention
Riccio, Dan, 204
Ridley, Matt, 128
Ries, Eric, 184
Ring
    approach of, 27, 35
    brand efforts of, 19, 201
    culture of, 206
    evolution of, 78–79, 88–89
    identifying sources of awareness of, 177
    marketing of, 26
    narrative of, 22
    overview of, 13–16
    product as marketing and, 164
    trust in, 166
Rivian, 35, 51–53, 55, 56, 64–65, 82, 128–129,
    201
Robinhood, 35, 83, 137, 144–147, 148, 163,
    166, 186
RxPass, 85–86

S

sales activation strategy, 170–171
scaling
    amplifying organic growth through
        community and advocacy for,
        161–164, 167
    brand power and, 147–148
    building trust as ultimate growth
        multiplier for, 165–166, 167
    campaign approach in, 155–156, 167
    challenges of, 202
    driving growth in, 148–151, 167
    introduction to, 143
    in launching, 139, 140
    premature, 124
    product as marketing for, 164
    public relations and, 162–163
    Robinhood's story of, 144–147
    staying true during, 186
    success strategy for, 149–150
    as too big, 213
Scaringe, RJ, 52–53
sequenced growth approach, 122–127
Shark Tank (TV show), 13, 14, 19
Shopify, 171

short-term metrics, 156
Siminoff, Jamie, 14–16
Sketch, 192
Slack, 39
smartphones, 20, 38, 82. *See also* iPhone
   (Apple)
Snapchat, 89
social media, positioning strategy of, 63
Spotify, 82, 89, 160–161
startups, 21, 88–89, 128, 148–149
Stephenson, Dave, 154
storytelling, 22, 79–84, 93, 106–109, 115,
   134–138, 141
streaming services, 60–61, 89–90
Stripe, 83
Structural Equation Model study, 155–156
superfans, 163
sustainable growth, 46, 146–147, 172–173,
   187. *See also* growth

T
target audience, 54, 73, 124, 132, 138
technology
   brand value of, 27–28
   capabilities of, 58
   to meaningful products, 21–23
   over customer needs, 58–59
   product experience as brand experience
      in, 20
   relationship transformation through,
      165
   as tool, 208
   universal challenge in, 171–173
   without purpose, 211–213
Tenev, Vlad, 144, 145, 146
Tesla, 51–52, 55, 56
testing, incremental, 185–186
test-learn-scale approach, 184
Thiel, Peter, 84
touchpoints, 41, 135, 158
trade-offs, in brand building, 31–33, 214
trust, 105–106, 147, 165–166, 167, 212–213

Turner, David, 86–87
Twitter (X), 63, 85, 121
typography, 104

U
Uber, 89, 205–206
underdelivering, 105–106
upper-funnel marketing, 177, 185, 187–188,
   221. *See also* marketing
user experience (UX), 38–39, 42, 47, 99, 107,
   222
   See also customer experience (CX); product
      experience
user generated content (UGC), 164

V
value delivery, defined, 147
Vega, Dave, 193
Verganti, Roberto, 78
vision
      as balancing with reality, 83–84, 93
      design principles and, 102–104, 115
      execution of, 104–106, 115
      misalignment of, 213
      shared, 102–104, 115
      through storytelling, 80–84, 93
      in workplace culture, 191
visual identity, 80

W
Wallace, Evan, 192
Waymo One, 126–127, 137, 165–166
WeWork, 83–84
why, for brand identity, 81–82
word of mouth, power of, 38, 127
working backwards process, 66, 196

X
Xiaomi, 82

Z
Zhao, Ivan, 120, 122, 198

About the Author

Lifang He has nearly 20 years of experience driving brand strategy, product marketing, and go-to-market excellence for some of the world's most influential tech companies, including Apple, Amazon, and Ring. From launching iPhones globally to pioneering entirely new product categories at Amazon, her career has been defined by building products and brands that scale fast and deliver results. Her work spans the full innovation life cycle, from zero-to-one launches to multibillion-dollar, category-defining businesses. She has earned industry recognition, such as a Cannes Lions Grand Prix Award, a Fastest Growing Brands Award, and a Jay Chiat Award for Strategic Excellence. She now leads a consultancy specializing in brand strategy, product innovation, and go-to-market.